"The civilized world stands appalled."
The New York Herald

AF800217

PUBLISHERS' NOTE

No cataclysm, not even the French Revolution or the Napoleonic Wars, nor earthquake or any other force of nature has so convulsed the whole current of human life and thought, or devastated property and destroyed life as has the great war which rages in Europe and affects a large part of the rest of the civilized world to-day. The effects of this war are now felt, consciously or unconsciously, by every man, woman, and child, and its results will influence and have direct bearing on the lives of unborn millions.

No event so momentous in its results or so far reaching in its effects has ever been recorded in the history of mankind. And no other event in the world's history has excited such tense interest as has the great drama of life and death now being played with the earth as a stage and all the millions of the civilized globe as spectators. It is only natural that mankind craves now, and will crave for years and centuries to come, for a history that will tell why this great catastrophe has come about, how it proceeded, and what its results. Of course, the whole truth about it will not be known for years; tens of thousands of volumes will be written about it, and a hundred years from now new facts will be discovered, just as to-day we are learning new things about our Revolution, and no one now living can expect to hear the last word on the subject.

But no intelligent person will, for this reason, be content to go without having the best account that it is possible to write with the data and knowledge that are available at present. We are of the opinion that every such person would be glad to have such a history for reading and reference *now:* as a result of this belief we have in course of preparation a series of volumes which we think will appeal to all thoughtful people.

There has been no effort to make haste in the preparation or issue of this history; the whole endeavor of the authors has been to gain perspective that would permit scholarly treatment, be as unbiased and non-partisan as possible, and of enduring value. The series is so planned that the early volumes form a very full and essential introduction to the record of the actual hostilities and enable the reader to grasp clearly the matters vital to a proper appreciation of the questions at issue and form a basis for a correct understanding of what this war actually is.

The elaborate introduction by ex-President Taft is a masterly and statesmanlike analysis of the causes and possible results of this great war and of America's position in regard to it.

The authors are George H. Allen, Ph.D., of the University of Pennsylvania, History Department; Henry C. Whitehead, captain in the United States Army; and Admiral F. E. Chadwick, U.S.N.

Dr. Allen, who is a recognized authority on the subject of European history and a writer on military affairs, is particularly well equipped for the task, for his investigations in modern European history have been made during long continued periods of residence and travel in Europe. He has lived four years in Germany; has made four journeys in Russia; has made a particular study of conditions in the Balkans, having had the special advantage of a meeting with King Nicholas of Montenegro. He has also studied the "Near Eastern Question" at Constantinople.

He was in Moscow when hostilities began, and as an American-born citizen of the United States was able to discuss conditions freely with officials. His various residences and travels in Europe while connected with the American Academy in Rome and later with the University of Cincinnati and as university extension lecturer in Berlin have been of invaluable assistance in the writing of this book. His work is a calm, judicial analysis, free from bias and prejudice: he stands as a historian should, aloof from the passions aroused since the outbreak of hostilities, and his sole object is to arrive at the truth. He makes plain many things obscure to the average person: for example, he shows how the struggle for the control of the Dardanelles and the "Near East" is one of the most important pivotal points and that the result of this struggle may be the final controlling factor in ending the war. Captain Whitehead is especially qualified for the reason that he has served in Europe, by official assignment, for observation. Admiral Chadwick's qualifications for treating of naval affairs are beyond question. He is not only a distinguished officer of high rank but also ranks high as an author. His career has been remarkable: he was naval attaché at the American Embassy, London; he has been Chief of the Intelligence Bureau and of the Bureau of Equipment; he was Chief of Staff of Admiral Sampson during the war with Spain and was advanced five numbers for eminent and conspicuous conduct in battle; he has been Commander-in-Chief of the South Atlantic Squadron and President of the Naval War College at Newport.

The first volume is devoted to a consideration of the events and conditions forming the potential and positive causes of the war and contains a well tempered and non-partisan examination into the motives that have led to this great conflict. It shows us that the *Welt-politik,* or general policy of Germany, was not, from Germany's point of view, fundamentally unjustifiable, and that the war between the western powers is only incidentally a commercial war; that Great Britain and Germany would not have gone to war on their own impulse; that aside from the invasion of Belgium, which was the positive, direct, and formal cause for Great Britain's participation in the war, there was only the apprehension caused by the sudden growth of the German navy as a potential cause; it proves that commercial rivalry was only very remotely connected with the various reasons for the war and that it ought not to be regarded as a cause at all; that the war so commonly regarded as one between Great Britain and Germany for supremacy in the west grew out of the inevitable clash between the Slav and the Teuton in the Balkans; and that in the Balkan peninsula existed the critical situation out of which the conflict necessarily emerged.

The second volume will be devoted to a review of the growth of the moral or spiritual factors which prepared the peoples' minds for war and brought the nations to the point of going to war, the preparation for war, offensive and defensive, in a physical or material way,—their resources in men, material, and money, etc., etc., and their mobilization,—indispensable to a proper understanding of the military and naval operations during the hostilities, and of immense value for reference when reading the later volumes and news of the day.

The third volume will contain a very full record of the outbreak of hostilities, the attacks on Serbia, the invasion of Belgium, and the advance toward Paris. Other volumes of the series will give accounts of the varied definite phases of the hostilities which the war assumes, and of the final results. Each volume will be complete in itself and be a full and satisfying treatment of its subject.

THIRD EDITION

REVISED

WITH ADDENDA

THE GREAT WAR

FIRST VOLUME
CAUSES OF AND
MOTIVES FOR

TO

W. M. A.

FRANZ JOSEPH CARL

FRANCIS JOSEPH I

Emperor of Austria and King of Hungary.

THE GREAT WAR

FIRST VOLUME
CAUSES OF AND
MOTIVES FOR

BY

GEORGE H. ALLEN, PH. D., OF THE UNIVERSITY OF PENNSYLVANIA, HISTORY DEPARTMENT; FELLOW IN CLASSICAL ARCHÆOLOGY, AMERICAN SCHOOL OF CLASSICAL STUDIES, ROME, ETC.,

WITH AN INTRODUCTION BY
THE HONORABLE WILLIAM HOWARD TAFT.

PRINTED *AND* PUBLISHED *BY*
GEORGE BARRIE'S SONS
at PHILADELPHIA

AMONG THE ILLUSTRATIONS IN THIS VOLUME ARE REPRODUCTIONS OF PHOTOGRAPHS COPYRIGHTED BY UNDERWOOD AND UNDERWOOD, BY PAUL THOMPSON, AND BY THE INTERNATIONAL NEWS SERVICE COMPANY

CONTENTS

LIST OF ILLUSTRATIONS

INTRODUCTION

Contemporaneous history is much less easy to study and learn than that of a preceding century. The difficulty is that the sources available to the ordinary reader are confined to fugitive magazine articles, or to the files of the daily press. The succession of events, as we gather them from newspaper reading from day to day, does not give to us a due sense of proportion as to their importance. This is true of the social, political, and economic record of our own country, and still more of the annals of foreign countries. When a cataclysm like the present European war makes a turning point in human progress and we are led to study its causes, with a view to a forecast of its effects, we find that if we have not been constantly and intimately engaged in a study of the international relations of the world we need the aid of a comprehensive summary of what has been going on in each country engaged in the war for a generation or more past: what have been its ambitions, its policies, the racial prejudices of its people, and where have been the points of conflict between its purposes and those of its antagonists. It may be conceded that an interval of at least half a century ought to elapse before history may be accurately, carefully, and judicially written. Then the heat of the issues has passed away, partisan controversy has faded out, and the post mortem evidence of witnesses, reluctant while in life, furnishes sources of information that are closed to the investigator *dum fervet opus.* Such a lapse of time, too, gives an opportunity to the historian to judge of the significance of events in the light of their results. In contemporaneous history we cannot avoid dealing with the future, because we must attach to the events we discuss an importance that is affected by our anticipation of their effect. Notwithstanding, however, the advantage that the historian who looks back fifty years or more has in painting a true picture of the life of a nation or the world, one who undertakes to recount, in their proper bearing, events of the current generation, serves a most useful purpose, aids much the future historians, and gives to one still taking part in the drama of nations a means of reviewing recent occurrences and summing them up as a useful predicate for future action. This is what Dr. Allen has done in the first volume of this series which is intended to deal with the great European war now absorbing our attention. He has divided the causes of the war into those which were potential and those which were immediate. Among the potential, he reckons racial prejudice and hostility and rivalry, and racial attraction and sympathy, and those irritating limitations of a nation's progress and growth under peace which furnish a motive for war. He rapidly reviews the history of each belligerent nation in the last half century, noting its industrial and economic growth, the changes in its political and governmental character, and its domestic and foreign policy. He thus brings out the conflicts of prejudice and interest between the nations, as well as the offensive and defensive alliances entered into for the purpose of maintaining peace and a balance of power, but which in the end have expanded a quarrel arising in a remote corner of Europe into the greatest war of history.

Dr. Allen is a Doctor of Philosophy, and a trained historian in the History Department of the University of

Pennsylvania, and has had great opportunities for studying the European situation. He spent some years in Italy, Germany, England, and France; he made four journeys into Russia. He made a particular study of conditions in the Balkans and looked into the "Near Eastern Question" in Constantinople. He was in Moscow when the hostilities began. His residence and travels in Europe while connected with the American Academy in Rome and later with the University of Cincinnati and as a university extension lecturer in Berlin fit him for his task.

His is not a partisan book. Within the limits to which he is confined, of course he cannot give all the evidence on each side of each issue, but he states succinctly the claims of each contesting nation and its vindication of its own course and enables the reader himself to draw the proper inference. Dr. Allen says: "My impartiality is the equilibrium of competing sympathy, not the empty neutrality of indifference. All the belligerent countries, excepting Serbia and Japan, are known to me through direct observation. But the fondest, most intimate associations unite me with Italy, Germany, France, and Great Britain. . . . A war between these countries is a source to me of unspeakable sadness. It is a conflict in which I must lose, whoever is victor." Occasionally he expresses a conclusion with which the reader may not agree, but only after a fair review of the attitude of each party.

My general information of European international relations is not such as to make my judgment of this book valuable to earnest and exact students of current European history, but I have interest enough in the subject and general knowledge enough to assure the ordinary reader that this book is a very interesting résumé of exactly what one desiring to keep abreast of current events would wish to find ready to his hand.

Dr. Allen approaches the subject from the standpoint of an American, which makes his opinion useful to American readers. He says that in spite of a cosmopolitan experience, it is impossible for him to conceal two native prejudices, a predilection for democracy and an aversion for militarism. Our public is divided into three classes—those who are earnest partisans of the Allies, those who are earnest partisans of the Teutonic countries, and those who deplore the war as a halt in the progress of the world, but who are grateful to Providence that our isolated position has up to this time kept us out of the conflict. The last class hope that we may continue to occupy the position of mediator only, and are profoundly convinced of the wisdom of Washington's advice that we avoid entangling alliances with European and other foreign nations, and continue to avail ourselves of the inestimable advantage of our geographical remoteness.

Those who are most earnest in their support of the Allies maintain the view that popular government and individualism and the guaranties of Anglo-Saxon liberty are all at stake in this war, that the success of Germany will mean the enforcement of the idea that human happiness is dependent only on efficiency in the administration of the government of the world, and that in the attainment of this, morality of nations and immediate considerations of humanity must be ignored or sacrificed. Their view is that if Germany overcomes her antagonists in this war, even if we of the United States are not drawn into it now, we shall be involved later, that the doctrine that "might makes right" in the general interest of progress must come into conflict with our purposes and principles, and must bring to the arbitrament of war the issue between German absolutism and American democracy; that a clash will necessarily come over our Monroe Doctrine, intended to protect America against forcible appropriation of American territory and suppression of independent popular government, and therefore that the whole weight of American influence should now be thrown upon the side of the Allies.

The pro-German element of our people is not entirely confined to those who were born in Germany or whose fathers were, but these so-called German-Americans form the chief part of those who believe that the contest is really by Germany and Austria, fighting for their existence, for the maintenance of Teutonic progressive civilization against Slav retrogression, and for the freedom of the seas. They represent the struggle to be against Russia with her Tartaric and Slavic hordes, against England and her claim to the monopoly of the ocean, and against France as the bound ally of Russia.

The violations of international law with reference to the rights of neutrals upon land and sea by both England and Germany have been such as to furnish illustrations and arguments for each one of these views. Those who are in favor of the Allies point with great emphasis and force to the fact that Germany in the conduct of the war has exemplified the principle that to efficiency must be sacrificed all humane considerations. The instance of the *Lusitania,* and the hurrying into death of more than one hundred defenceless American citizens—men, women, and children—give much weight to this contention.

I believe that the great body of American people wish to avoid a conflict. They are anxious that the President should assert our rights as neutrals and the rights of our people as neutrals under the accepted principles of international law, and they are willing to go with him as far as he deems it wise to go, hoping that some solution may be found that will not make it necessary for us to range ourselves forcibly on the side of the Allies. Whether the persistence of Germany in her justification of the loss of life on the *Lusitania* will lead to a breach, it would be unwise to forecast. There is a course short of declaring war that the United States might take, and that is, to sever diplomatic relations and decline to have association with a country whose methods of warfare are so lacking in humanity. As long as the interference with trade is limited to an appropriation of property which may be recompensed in the future by a judgment of an arbitral tribunal, a protest may perhaps suffice; but where human lives of unarmed and defenceless Americans are taken and the unjust and inhumane policy is avowed which will involve the future taking of such lives, protest with the prospect of a mere claim and possible judgment for damages can hardly satisfy the national right and honor.

As I write, the issue is before the President and we are all awaiting the conclusion with intense interest. Of course the President may not declare war—that is given by the Constitution to Congress—but so strong is the patriotic sentiment, so intense the desire to stand behind the pilot of the ship whom the people have selected, that under prevailing conditions, it is after all the President who must decide. If he calls Congress together with an intimation that in his judgment drastic action is needed, we can be confident that Congress will follow the intimation. The closeness with which we have been brought to the European conflict and the really critical situation in which we find ourselves make it important that our public should have clearly in mind just what the issues are upon which this European war has turned.

The Napoleonic wars, of course, were world wide in their influence, and affected the United States seriously. But the progress since then made in the machinery of war, in the range of artillery, in the resisting power of armor, in the use of the submarines and aëroplanes, in the destructive effect of high explosives, and in the speed of marine transportation, create conditions that make the ocean much less of a defence for us than then, and bring us much closer to the actual conflict. Those of us who are optimists are hopeful that the inevitable prostration of even the victors in the present controversy, due to the awful waste of men and treasure, will make the anticipated danger to our peace remote. In any view, the importance of keeping ourselves advised of the moving causes of the war and their effects is manifest.

The unification of Germany as the direct result of the Franco-Prussian War furnishes a good starting point for the story of the war. The ultimate failure of the present Emperor to maintain the Triple Alliance and at the same time preserve a friendly intimacy with Russia, as advised by his grandfather, led to the Dual Alliance between Russia and France. In spite of the Berlin Congress and its impossible settlement "with Peace and Honor" of the Near Eastern Question, Mr. Gladstone's campaign, founded on the Bulgarian atrocities, and his unreconcilable antipathy for the Turks, led gradually to the substitution of Germany for England as the Friend of the Sick Man of Europe. Germany's consequent activity in the promotion of enterprises in Asia Minor and the construction by German capital, under Turkish concession, of the Bagdad railway, formed a line from Germany toward Persia which crossed the Dardanelles and Russia's outlet to the Mediterranean, and formed a probable route to India, midway between the northern land route from Russia, through Afghanistan and Persia, and England's route by the Suez Canal. This, with other circumstances, led to a common policy between England and Russia in respect to

Afghanistan and Persia and the northern railway to India and to precautionary steps to prevent the Bagdad railway plan from prejudicing English and Russian interests. The Boer War, the fateful telegram of Emperor William, and the new naval policy of Germany, aroused the fears of England and prompted renewed activity on her part in naval construction. From this can be traced a growing suspicion and strain between the two peoples of these nations.

The evident understanding between Russia and Germany as to the Japanese war indicated a good feeling between them for the time being, which was not, however, strengthened by Austria's annexing Bosnia and Herzegovina, with Germany's approval, after Russia's defeat.

Dr. Allen describes the remarkable congeries of peoples under the Hapsburgs in the dual Monarchy of Austria-Hungary. The Germans of Austria, the Magyars of Hungary, the Czechs of Bohemia, the Poles of Galicia, the Roumanians of Transylvania, the Serbian Slavs of Croatia, the Slavs of Bosnia and Herzegovina, and the Italians of Trentino, Trieste, and Istria, have been here united in a curious mosaic. The Poles were given a milder and more enlightened government under Imperial Austria than their brethren of Russia or Germany. The Slavs of Croatia, a dependency of Hungary, however, were treated with less liberality by the ruling Magyars and their racial brethren of Bosnia and Herzegovina, while enjoying a good government, also were kept under rigid control, due to the dangerous proximity and sympathy of Serbia. The Czechs of Bohemia, though Slavs, enjoy a more liberal autonomy and have greater influence in the Empire. The Imperial Government in its Italian territory has felt it necessary to cross the desires of its Italian subjects in language, education, and other matters.

That such an unnatural union could be maintained under one government is most noteworthy, and the difficulties of its maintenance fully explain the absence of uniform treatment of the different elements of the whole. The impatience and discontent of some of these peoples of the Austrian Empire at not having greater voice in the Imperial Government, and the restlessness of others at being under that government at all, have produced a condition not making for quiet in southeastern Europe. The recalcitrant subjects of the Hapsburgs found encouragement for their separatist aspirations in the neighboring nations whose peoples were of the same race and language. The Slavs looked to Russia and the Russian people as their friends and protectors. When Bosnia and Herzegovina under the suzerainty of Austria, as provided by the Berlin Congress, were annexed by Austria, the change in actual government was not a great one, but taken as it was, with Germany's approval, when Russia had her hands tied by the consequences of the Japanese war, as already said, it left a scar. It injured Russia's prestige in the Balkans with her Slavic wards and made her more sensitive than ever to future aggression by Austria.

The Morocco question forms another turning point, when Germany by thrusting herself forward to obstruct France's colonial expansion, in which Germany had little or no interest, manifested a spirit that did not bode good for future peace and had the effect of definitely ranging England among the friends of France and of confirming the *Entente Cordiale.* The strained feeling between Germany, on the one hand, and England and France, on the other, however, seemed to have abated after the Algeciras Treaty, and all were hopeful that danger of a breach was past. While this calm was reassuring, the international grouping of Russia, France, and England against possible German aggression was an accomplished fact, and was to be reckoned with, should occasion arise for its manifestation.

Just before the outbreak of the War, Russia was disturbed with great strikes indicating political unrest; France seemed much torn with socialistic resistance to her military preparations; and England was busy with the threatened rebellion in Ulster over Home Rule for Ireland. German diplomatists were unable to put themselves into a Russian, French, or English frame of mind and correctly to estimate the real effect that such internal dissensions might have on the fighting spirit of either nation. They evidently deemed the three nations quite unprepared for war, in view of these domestic weaknesses. In this respect, German diplomacy was certainly at fault, though few could have anticipated the magnificent spirit of patriotism, the calm dignified determination, and the silent courage with which the French people have faced this awful trial.

In the First Balkan War, much of Turkey's territory was taken by the Balkan Allies. This was followed by the ineffective Congress at London. Then came the Second Balkan War over the spoils, in which Bulgaria was humiliated by Serbia, Greece, and Roumania. These wars kept Russia and Austria in feverish excitement and diplomatic activity. Austria was determined to prevent Serbia's getting to the Aegean Sea at Salonica, and did so.

Looking back over these rapidly succeeding events, and considering the clashing interests as they are brought out clearly, succinctly, and fairly by Dr. Allen, we can easily see what a Pandora's box the Balkans were and how the great war grew out of its opening.

The murder of the Austrian Crown Prince and his wife, as the work of a pro-Serbian conspiracy, and Austria's ultimatum to Serbia, set a spark to the powder that we all must now recognize as covering the whole of Europe. The Alliance and the Entente dragged all the powers in, and the train of circumstances leading to the War is as clear as possible, in the event.

The invasion of Belgium and the avowed breach of international obligation by Germany requires from the author a summary of the history of Belgium's neutrality guaranteed by all the Powers, including Prussia and Austria, and its reaffirmation by the North German Confederation for the period of the Franco-Prussian War. This breach of a sacred treaty obligation forms the basis of a capital indictment on which Germany will certainly be convicted at the bar of history.

The subsequent joining of the Allies by Japan and Italy, and of the Teutonic Alliance by Turkey, calls for a chapter upon these three countries and a rapid résumé of their recent history in its bearing upon their interests in the conflict. Japan's alliance with England and her desire to deal with China when the European powers could not intervene, Italy's insistence on the balance of power guaranteed her in the Adriatic by her treaty with Austria, and her wish to add to United Italy the Italians of the Trentino, Trieste, and Istria, and Turkey's business association with, and friendship for, Germany, under the influence of the Young Turks, with Enver Pasha at the head, and her natural antagonism for Russia, furnish the respective motives for their coming into the fight.

Dr. Allen discusses the evidence as to the immediate responsibility for the war, and considers the charge that the bringing on the war was of a deliberate purpose by Austria and Germany. He refers to the remarkable coincidence of the completion of Germany's preparation for war in the finishing of the Kiel Canal; the very great enlargement of her military forces in the years of 1913 and 1914; and her construction of a remarkable network of railways, strategic but non-commercial, on the Belgian border, as circumstances tending to show that Germany had the war in mind. The possible explanation of this he offers is that the confirmation of the *Entente Cordiale* really convinced Germany that the war was likely, and that she must perfect preparations to meet it. He points out that the murder of the Heir Apparent, which was the immediate occasion for the issue, could not have been a matter of premeditation, and then shows by diplomatic correspondence that Germany and Austria really thought that Russia was in such a condition of unpreparedness and internal dissension, a matter to which I have already referred, that Austria's demands in respect to Serbia would be acquiesced in. He finds Russia's persistent mobilization at the critical juncture a moving cause for the breach, and attributes that to the military party in Russia, to which he thinks the Tsar was opposed. He credits the Tsar with an earnest desire for peace.

It is difficult to escape the conclusion that had Germany desired peace she might have secured it by inducing Austria to hold her hand in attacking Serbia until a conference of the Great Powers, or of those not immediately interested, had suggested "a way out." Serbia seemed willing to conform to most of Austria's demands and to be anxious to do anything, not involving complete surrender of her independence, and Russia was not disposed to prevent this. The lack of preparedness on the part of all the Allies is the strongest evidence of their wish to avoid a war. Probably Germany felt that a war must come in the near future, and that, with the odds against her and Austria in ultimate numbers and resources, now was the best time, when she was in the pink of preparation and her adversaries were not.

Our author considers the German *Welt-politik* and expresses some sympathy with Germany's impatience and protest against the appropriation by all the other nations of the face of the earth before her great industrial and political development and growth began. He likens the situation to the ownership by a few of the large estates in England and Ireland, and points to the policy to which England was forced in Ireland of ignoring private property rights and of compelling a redistribution. This is not the place to consider the other side of such a view, or the force of the analogy, which I would be much disposed to question.

On the whole the volume is a most useful one for a practical survey of the European situation leading to the War, and bears on every page evidence of the impartial spirit in which it is written.

If the other volumes, *i. e.*, a second on the military preparedness of each belligerent country, a third on the diplomatic preliminaries, the mobilizations, the attacks on Serbia and the invasion of Belgium, and others on the progress of the War, are as clear, concise and helpful as the one now published, they will constitute a real addition to the literature on this great World disaster.

WILLIAM H. TAFT.

PREFACE

............licet mihi praefari quod in principio summae totius professi plerique sunt rerum scriptores, bellum maxime omnium memorabile, quae umquam gesta sint, me scripturum...........Nam neque validiores opibus ullae inter se civitates gentesque contulerunt arma, neque his ipsis tantum umquam virium aut roboris fuit........... Livy 21, 1.

[*I may be allowed to prefix a remark, which most writers of history make in the beginning of their performance, that I am going to write of a war, the most memorable of all that were ever waged..........For never did any other states and nations, of more potent strength and resources, engage in a contest of arms, nor did these same nations, at any other period, possess so great a degree of power and strength.*]

Nations, like individuals, pass through periods of crisis, when emotion and moral conflict are so intense that the experience of generations seems to be crowded into as many days. The current of human life was suddenly engulfed in one of these whirlpools of excitement and anxiety during the midsummer days of 1914, when the bewildering confusion of contrasted sensations, hope and despair, enthusiasm and rancor, impressed their vivid and ineffaceable mark upon the memory of those whose lives and happiness were involved in the momentous decisions which were in suspense.

In times of tranquillity the imagination dwells with curiosity upon the crucial periods of the past and strives to conceive them as real and to derive from them definite impressions as of contemporary events. But as soon as our exertions are relaxed, they pass again into a nebulous state, leaving us with a lurking, half-conscious feeling that such sensational epochs are on the remote borderland of reality and will not return.

But almost without warning we were brought face to face with such a tremendous situation as our habitual thoughts and feelings had associated with another world of events, one entirely foreign to our own lives. The imagination was baffled, dazed, paralyzed, before the reality of what it had often pondered over as an abstraction.

As I wandered through the pleasant, shaded streets of the Petrovsky suburb of Moscow, on Sunday, August 2nd, the strollers whom I encountered showed the peaceful unconcern of the weekly repose. On the verandahs, or through the open windows of homelike villas, one caught glimpses of family groups about the steaming samovars. An air of undisturbed tranquillity was everywhere in evidence. I almost persuaded myself that, after all, the perturbation of the week just passed and the alarming despatches announcing warfare were a troubled dream which had departed. But the view from the bridge at the station of the Warsaw railway brought back the grim reality. Long trains of freight cars under military supervision filled the tracks as far as the eye could reach, and these were being loaded with all kinds of munitions and supplies which had been brought and deposited wherever possible along the line. Constant processions of vehicles and commandeered horses were arriving from all directions. The scene of feverish activity contrasted harshly with the Sunday calm. Mobilization—the word to conjure with, the word that loomed vaguely but ominously in our imagination—mobilization was in full progress.

The absorbing, overpowering feature of the great war is its dramatic character; the startling suddenness of its approach, the irrepressible rapidity of its extension, and, above all, the reckless, unhesitating determination with which interests of unmeasurable magnitude were staked upon the uncertain issue.

It is true that a definitive history of the war is impossible until we can view the events with calmer detachment and truer perspective, after they have dropped well astern in the wake of the progress of time. But it is just as truly incumbent upon us to correlate the events of the mighty present, and to endeavor to understand, as adequately as our limited perception will allow, the forces in action which are reshaping the framework of society for future generations of mankind.

The succession of circumstances and an ever increasing interest, stimulated by long residence in Europe, in the currents of politics and society which were converging towards the maelstrom have induced me to undertake the composition of the present work. Amid the myriad of books which have appeared even in this country dealing with the background and actual course of the war, it would be vain to claim for this work any unique characteristics. But whatever usefulness it may prove to possess is based chiefly upon its comprehensive scope and impartial treatment.

The more salient elements in the narrative are not likely to be greatly altered by future investigation. It has been my chief aim to provide a clear, unbiased account of these facts, to serve as a reliable foundation upon which the thoughtful reader may build, as time will reveal new data and points of vision. My impartiality is the equilibrium of competing sympathy, not the empty neutrality of indifference. All the belligerent countries excepting Serbia and Japan are known to me through direct observation. But the fondest, most intimate, associations unite me with Italy, Germany, France, and Great Britain. Their fields and forests, highways and byways, cities and villages, have become part of my life. A war between these countries is a source to me of unspeakable sadness. It is a conflict in which I must lose, whoever is the victor.

But in spite of a cosmopolitan experience it is impossible for me to conceal two native prejudices, a predilection for democracy and an aversion for militarism.

Much of the material in the first volume is the gradual accumulation of travel and enquiry. An insoluble debt is due to German scholarship for the facilities which have always been cheerfully offered, and above all, for the inspiration and instruction of investigators in their scientific methods of work.

The current periodical literature and daily press of the leading countries, the official correspondence and reports, and a large part of the special literature on the war have been brought into requisition for the present work; while the Annual Register, Schulthess' *Geschichtskalendar*, and *L' Année politique* have been of invaluable assistance in the verification of dates and in supplying the texts of speeches and the more important parliamentary proceedings.

GEORGE H. ALLEN, PH.D.

History Department,
University of Pennsylvania.

Beaconsfield. Andrássy. Bismarck. Schuwaloff. Salisbury.

The Congress of Berlin, 1878. *After the painting by Anton von Werner. The delegates met on the 13th of June and on July 13th signed the Treaty of Berlin, which changed the boundaries of the Balkan States from those outlined by the Treaty of San Stefano.*

CHAPTER I

Motives and Forces in the Territorial Adjustment of European States

The present situation justifies a retrospect. The most powerful factor in moulding European states. Theory of nationalities. The Unification of Italy. *Italia Irredenta;* Irredentism. Unification of Germany. Bismarck; *Real-politik.* Exceptions to the application of the Principle of Nationalities; Alsace-Lorraine; Examples in the Russian Empire; prejudiced opinion of Russian expansion; Finland, the Baltic Provinces, and Poland. Hapsburg Realm apparently the Negation of the Principle of Nationalities. Why a dual, not a federal monarchy. Races in Austria-Hungary; Slavs the most numerous. Poles in Austria, and in Prussia. Balance of Power. Standing alliances. Origin of the Triple and Dual Alliances.

It is surely no more than a modest assertion to predict that the Great War, which bewilders the intellect with its magnitude and horror, will mark the beginning of a new historical epoch. The governments of more than half the inhabitants of the world are at present engaged in the struggle. Our attention is enthralled as by a drama enacted on a world-wide stage. Our imagination is alert to grasp some indication upon which to build a conjecture of the coming age. The present is not a propitious time for tranquil retrospection. And yet the key to the problems of the present and future lies in the immediate past, in the period which has just reached its termination. The old century, likewise, was ushered in with the turmoil of general warfare, and through it run the tendencies and forces which converge upon the present crisis. A brief review of some of the significant features of the last century will lead naturally to a discovery of the political, social, and industrial movements, and the national ambitions and policies, of which the more immediate causes of the present conflict are the consequence.

If we consider the nature of causes of war, we shall perceive that they are capable of division into two great classes, which for convenience we may agree at the outset to call *potential* and *positive* causes respectively. Potential causes will include all institutions, tendencies, movements, contrasts, in short all features or elements of social life and organization, which are a fertile field for implacable discords. The class of positive causes will be made up of the actual situations, disagreements, collisions of interest, of which the state of hostilities is the direct consequence. The potential causes are the immanent, inherent, intrinsic, or general causes. The positive are the determinate, definite, explicit, or special causes. Examples as illustration of the distinction between the two kinds of causes readily suggest themselves. The potential cause of the war which resulted in the independence of the United States was the remoteness of the American colonies from the mother country. The positive cause was the question of taxation. The potential cause of the Civil War in the United States was slavery. The positive cause was the assumption on the part of certain of the states of the right to withdraw from the Union. The resumption of hostilities by Great Britain and France in 1803 had as potential cause the insatiable ambition of Napoleon. Its positive cause was the non-fulfilment by Great Britain of the terms of the Treaty of Amiens. The desire for Italian unity was the potential cause of the campaign of 1859, the intervention of Napoleon III in the affairs of Italy was the positive cause.

In the following brief summary of the most fruitful political movements in the nineteenth century in Europe, and of the recent condition of affairs in the principal states, the causes of the war as such will not be directly discussed. But by choosing for treatment from the vast field that lies before us the tendencies that have an enduring significance these introductory observations will be made suggestive of the potential causes, although not explicitly indicative of them.

To understand the problems that attend the progress of European states by reason of their mutual relationship, we must consider before all two political ideals of fundamental importance. These are the Principle of the Balance of Power, and the Principle of Nationalities. One is the rather evasive goal of diplomacy, the other the incentive of popular enthusiasm. The Balance of Power was a guiding motive in foreign policies of European states before the present conception of a nation had gained recognition. States were frequently recast in the eighteenth century, or bartered to and fro, to establish an equilibrium of power, or provide satisfaction for rival dynastic claims, without any thought for the feelings of their inhabitants, as though they were the estates of great proprietors. However, the Principle of Nationalities created in the nineteenth century several of the most important states, which are the units with which the Balance of Power is constructed. It is convenient, therefore, to treat the younger principle first.

In the French Revolution a whole people came to consciousness of its national solidarity, its mission, its ideals and aspirations, with the sudden fury of a mighty volcanic eruption and the resulting disturbances have vibrated throughout the nineteenth century, developing greater violence from time to time when it has been necessary to disengage weightier masses of political inertia and oppression. The development of national self-consciousness in the peoples of Europe was contagious. It made the Doctrine of Nationalities a live principle. A state and a nation are not necessarily the same. A state is a conventional political organization, a unit of sovereignty. A nation is a living organism, the result of a gradual process of growth. It is bound together by the most intimate ties, a common language and literature, common customs, and a common consciousness of right and wrong.

Formerly the system of states disregarded national divisions. Boundaries were frequently traced with supreme indifference to national distinctions, not to mention national preferences. It is a singular fact that the nineteenth century has witnessed in Europe concurrently the rapid advance in means of communication, which tend to break down barriers and foster the growth of cosmopolitanism, and the remarkable revival of the spirit and consciousness of individual nationalities. The increased facilities for study and investigation and the cheapening and general diffusion of books probably contributed to the awakening of this national spirit.

The Principle of Nationalities became an impelling force and one which continues to demand consideration. It affirms that states and nations should correspond; that each nation should have its own exclusive, independent political organization; that sovereignties and nationalities should be co-extensive. The nation is the proper social foundation for the state. The national state is the most modern product of political development. It affords the most favorable field for the growth of free institutions. It diminishes or eliminates the friction between the central and local organs of government. The states with a composite population, where national unity does not exist, have lagged behind in the progress of the nineteenth century toward popular government.

The most conspicuous achievements in consummating the popular yearning for national solidarity have been the unification of Italy and Germany. These movements have been central features in the political developments of the nineteenth century.

Except for a few years under Napoleon, Italy was reckoned

ITALY
IN THE
XIX CENTURY
Scale of Miles
0 50 100
The dates are those of the recognition by the various provinces of the King of Sardinia
AUSTRIAN EMPIRE
VENETIA
1866
Verona
Venice
Villafranca
Solferino
LOMBARDY
Magenta
1859
Milan
Novara
Vercelli
Turin
Susa
Cavour
SAVOY
Rhone
FRANCE
KINGDOM OF SARDINIA
PIEDMONT
LIGURIA
Genoa
Nice
Monaco
(DUCHY)
Parma
Reggio
(DUCHY)
Modena
Finale
Ferrara
Ravenna
Bologna
ROMAGNA
1860
Lucca
Pisa
Florence
Leghorn
GRAND DUCHY
Siena
OF TUSCANY
1860
ELBA
Pesaro
Ancona
THE MARCHES
1860
PAPAL
STATES
UMBRIA
1860
Rieti
THE PATRIMONY
1870
Rome
Chieti
CORSICA
(To France)
SARDINIA
Cagliari
ISTRIA
DALMATIA
ADRIATIC SEA
Foggia
Benevento
Capua
Gaeta
Naples
Castellamare
Bari
Brindisi
Taranto
BASILICATA
Otranto
Gulf of Taranto
TYRRHENIAN SEA
KINGDOM OF THE TWO SICILIES
CALABRIA
Milazzo
Reggio
Palermo
Messina
Termini
Calatafimi
SICILY
1860
Catania
MEDITERRANEAN SEA
AFRICA
8° 10° 12° 14° 16° 18°
46° 44° 42° 40° 38° 36°
Longitude 12° East from 14° Greenwich

as merely a geographical expression in the early part of the nineteenth century. Though no larger in area than Arizona, Italy was divided between ten distinct sovereignties. Piedmont alone in the northwest, including Liguria and the island of Sardinia, was ruled by a native dynasty. The Piedmontese were hardy, thrifty, conservative. Their institutions still exhibited vestiges of feudalism. The policy of the government was narrow and petty. The heavy atmosphere of Austrian absolutism depressed the heart and brain of Italy. Lombardy and Venetia, the richest part of the country, were under the direct authority of an Austrian viceroy. Bitterness against the foreigner has obscured the better features of Austrian rule. Order was maintained, and a fairly impartial administration of justice, excepting where political questions were involved. But the Austrian provinces bore an excessive burden of taxation, and all agitation for political reform was repressed with the utmost severity. Many eminent men were arrested without trial and languished for years in Austrian military prisons, their only crime having been their patriotism. The duchies of Parma, Modena, and Tuscany were governed by protégés of Austria, whose policy was usually directed from Vienna. The spirit of inertia and decay brooded over the Papal State, and the Neapolitans groaned under a Bourbon despotism, the most despicable tyranny in Europe, a government which Mr. Gladstone at a later period declared to be the negation of God. From such unpromising conditions Italy arose rejuvenated, purified by eager self-devotion to a common national aspiration.

Secret societies eluding a despotic and inquisitorial police system carried on a liberal agitation leading to the two revolutions of 1820 and 1831, which were barren in immediate results, but taught the people that liberty without unity was impossible. Then little Piedmont assumed her heroic task as leader in the conflict for unity. She defied the haughty oppressor of Lombardy and Venetia. Aided at first by contingents from the other states, whose people insisted upon co-operation, later deserted by them as the tide of revolution throughout Italy waned, the Piedmontese army was defeated at Custozza in 1848, and at Novara in 1849, and Piedmont had to make such terms of peace as she could. But she retained her liberal constitution, which had been proclaimed as the inauguration of her heroic policy, to serve as a beacon-light in the gloom that again spread over Italy.

Though crushed in war, the fortunes of Piedmont were revived and lifted to immortal renown through the genius of a single individual, the most remarkable diplomat of the nineteenth century, Count Cavour. By causing Piedmont to participate in the Crimean War, he obtained the right to speak in a European congress, where he exposed the wrongs of Italy, and brought about the alliance with Napoleon III and the latter's intervention against Austria in 1859. After her defeats at Magenta and Solferino Austria yielded Lombardy. The next year witnessed an exploit which may challenge comparison with the narratives of epic poetry. The national hero, Giuseppe Garibaldi, slipped out of the harbor of Genoa by night with about 1100 volunteers embarked in two steamers. Landing stealthily at Trapani near the western extremity of Sicily, May 11, 1860, he cleared the island within a few weeks of the 24,000 troops of the Bourbon king of Naples. He was everywhere received as a deliverer. He crossed to the mainland, united his forces with King Victor Emanuel and the Piedmontese army, which had traversed the Papal State to enter the field, and dealt a death-blow to the power of the Neapolitan Bourbons on the banks of the Volturno, Oct. 1. Meanwhile the duchies threw in their lot with Piedmont, so that Victor Emanuel ascended the throne of the Kingdom of Italy, March 17,

1861. The process of unification was continued by the annexation of Venetia as a result of the alliance with Prussia in 1866, and the occupation of Rome upon the withdrawal of the French garrison in 1870.

The Doctrine of Nationalities implies that a government should receive its sanction by the consent of the governed. For community of sentiment is the supreme test of common nationality. True to this principle the Italians submitted each of these annexations to the judgment of the populations concerned and in each instance union with Piedmont or the Kingdom of Italy was confirmed by a plebiscite on the basis of universal suffrage with an overwhelming majority. There remains an Italian population of several hundred thousand souls still included in the Austrian territory, in Southern Tyrol about Trent, in Trieste, and in Istria, all adjacent to the Kingdom of Italy; and despite the connection with Austria-Hungary through the Triple Alliance, the hope that these people may ultimately be united with the great mass of their Italian brethren has been cherished by many, who call these territories *Italia Irredenta*, or Unredeemed Italy. Those who refuse to admit that the Unification of Italy has been completely achieved because this population has remained outside are called Irredentists, and their attitude, Irredentism.

In the political development of the nineteenth century the liberal movement was less pronounced in Germany than in Italy. The forcible, warlike policy by which unity was finally achieved in Germany reduced the liberal movement to a subordinate, dependent position. The spirit of revolution was relatively less vehement because the rulers were native princes by descent and tradition and their administration was in general less oppressive. Dynasties and people had grown together by long association. This was at the same time an advantage, and a hindrance. The present rather limited constitutional regime was obtained with very little bloodshed and violence. But on the other hand, the spirit of individualism of the different states was too strong to be entirely eradicated. And so the united Germany became a federal state, not a unitary, constitutionally compact, state like Italy.

The ideal of a free, united Germany was born in the fervor of the national awakening when the armies of Napoleon were swept back into France. But in the disposition of affairs which followed in the Congress of Vienna, German patriots were cruelly disappointed in their hope of unity; liberals were deceived in their expectation of constitutional government. True, progress had been made in diminishing political chaos. In place of about four hundred states in Germany there were now less than forty. But these were joined in a very loose, inefficient confederation, with a diet meeting at Frankfort composed of the delegates of the various princes, without discretionary power, voting according to the instructions of those who sent them. The powers of the diet were chiefly negative, that is, repressive. Austria was always to have the presidency of it, although only one-third of the Austrian Empire was included within the bounds of the confederation. The Austrian policy of reaction dictated by Metternich predominated for more than a generation.

Each of the thirty-eight states comprising the confederation after 1815 had its own customs boundary as a barrier against the sister states as well as foreigners. To the different parts of Germany, economically interdependent by nature, this was an intolerable obstacle, and Prussia took the lead in establishing a tariff union in 1819, which by 1834 embraced seventeen states with an aggregate population of 23,000,000. This Zollverein, as they called it, was geographically a forecast of the present Empire. Thanks to this fiscal union the states were effectively knit together by commercial ties before the way was clear for political unity.

The whirlwind of revolution which swept over Europe in 1848 prostrated for a time the power of reaction in Germany. A national parliament elected by the people assembled at Frankfort, drew up a federal constitution and offered the imperial crown to Frederick William IV, King of Prussia; but he refused it. Prussia was still a docile follower of Austria.

The mid-century period of commotion in Germany, as in Italy, left no tangible results, except that Prussia, like Piedmont, emerged from it with a constitution. The parliamentary method of achieving the political unification of the country had signally failed. The Liberals hoped that Prussia would identify her policy with their ideals, and play the role of Piedmont in Germany. Prussia devoted all her energy to the unification of the Fatherland. But the initiative, the spirit, and the guiding force of the movement were henceforth monarchical. Moreover, Prussia was not merged in the united commonwealth losing her identity like Piedmont. She transformed in large measure the rest of Germany, which received the characteristic stamp of her institutions.

The failure to bring about unity by deliberation was clearly due to the tenacity with which the German rulers clung to their independence. The only practical means to overcome this difficulty was for one state to reduce the others to virtual subordination by force. Two states far exceeded all the others in power. But of these, Austria was poorly qualified to become the dominant member in a federal German union, because her spirit was not representative, and her attention was distracted by internal problems arising from her non-homogeneous population. Prussia was qualified by her character and traditions to assume the supremacy; and since it was hopeless to reduce Austria to a dependent position within the union, it followed that Austria must be excluded from it.

After her crushing defeats and humiliation by Napoleon, Prussia had submitted her antiquated institutions to a thorough overhauling and reorganization. The state was in large measure reconstructed by Hardenberg, Stein, and Scharnhorst. The administrative reforms of that time are only less important as models for continental European practice than those inspired directly by the French Revolution. The most conspicuous one was the establishment of the rule that all male subjects who are physically fit must serve for a limited period in the army. Prussia has faithfully upheld the theory of universal obligation to military service, although the rule has not at all times been thoroughly enforced. William I, who became regent in 1858 and sovereign three years later, perceived that German unity must be achieved by force, and accordingly insisted that the military establishment should be expanded so that all young men arriving at the military age could be admitted as recruits, and that the period of service in the active reserve should be doubled. The determination of his minister Prince Otto von Bismarck carried out this design in the face of the bitter opposition of the lower house of the parliament and in practical violation of the provisions of the constitution guaranteeing the control of the budget by the chamber. The results of this policy while appearing to justify the irregularity of the means by its preeminent success might easily have established a very dangerous precedent for constitutional government in Prussia.

Bismarck possessed a clear perception for reality, and an unusual quality of intellectual astuteness and subtlety. He was endowed with an iron will and tenacity of purpose. He was capable of conceiving the most audacious projects of statesmanship and had the capacity and resolution for bringing them to sucessful issue with startling precision and assurance. He was a representative of the sturdy Prussian landed aristocracy with many of their prejudices. He exemplified the harsher quality of the Prussian temperament, which is in harmony with the less genial soil and climate, and the stern conditions under which the Prussian people have asserted their claim to live and to expand. His disposition was in accord with the spirit of materialism which had mastered the temper of the age, substituting sober realism for idealism, prose for poetry, facts for fancies, and applied science for abstract speculation. He held in contempt the liberal ideas of 1848-9. He upheld the monarchical traditions of Prussia, and he steadfastly maintained that the English system of the parliamentary control of government was neither constitutional nor desirable in his own country.

We may appropriately dwell upon the character and aims of Bismarck. He left an enduring impression upon the institutions of Germany, where his spirit is still alive, and exerts a profound influence. His diplomacy was not free from criticism on ethical grounds. It was not infrequently characterized by artifice and duplicity, which are often excused as unavoidable. The unparalleled success of his diplomacy contains a possible element of danger, because it might commend his methods to later, and less capable, statesmen inclined to the belief that the end justifies the means. But Bismarck's spirit stands primarily for organized administrative efficiency, for a carefully trained,well disciplined bureaucracy, things that in an age of science and specialization may as justly claim our attention as the hallowed organs of popular and representative government.

Bismarck's career embodies a conception of the proper attitude and aim of the state in its relation to other states which is accepted by leading political philosophers in Germany, and called *Real-politik.* It is based upon the assumption that the chief aim of the state is power. It means the exclusion of ideal objects from foreign policy, and concentration upon material ends. But the exponents of the system urge a clear perception of the attainable and the strict exclusion of all that is impractical, visionary, or chimerical.

Bismarck's resolute policy is illustrated by a famous speech which he delivered in parliament, in 1863, in the course of which he declared; "Prussia must keep her strength intact for the favorable moment, which is too often missed. Prussia's boundaries are not favorable to the development of a strong body politic. Not through fine speeches and majority resolutions will the questions of the hour be decided—that was the mistake of 1848 and 1849—but by Blood and Iron". His "blood and iron" policy became proverbial. The stern temperament of her great leader and the struggles attending her rise impressed upon the new Germany a profound appreciation for strength, and the conviction that in the last resort a nation is based on force, and only by force can be held together.

The repeated blows of the mighty Prussian hammer forged in three wars within seven years the rivets that were to bind together the German Empire. Bismarck foresaw events with never-failing clearness of perception, and not the least proof of his ingenuity is his success in isolating the powers which were in succession the enemies of Prussia. Bismarck did not shrink from promoting conflicts which he knew to be unavoidable. The annexation of Schleswig by Denmark offered a pretext for the regulation of the unsatisfactory situation of the two duchies Schleswig and Holstein by Prussia and Austria together in 1864.

Bismarck manipulated the problems arising in connection

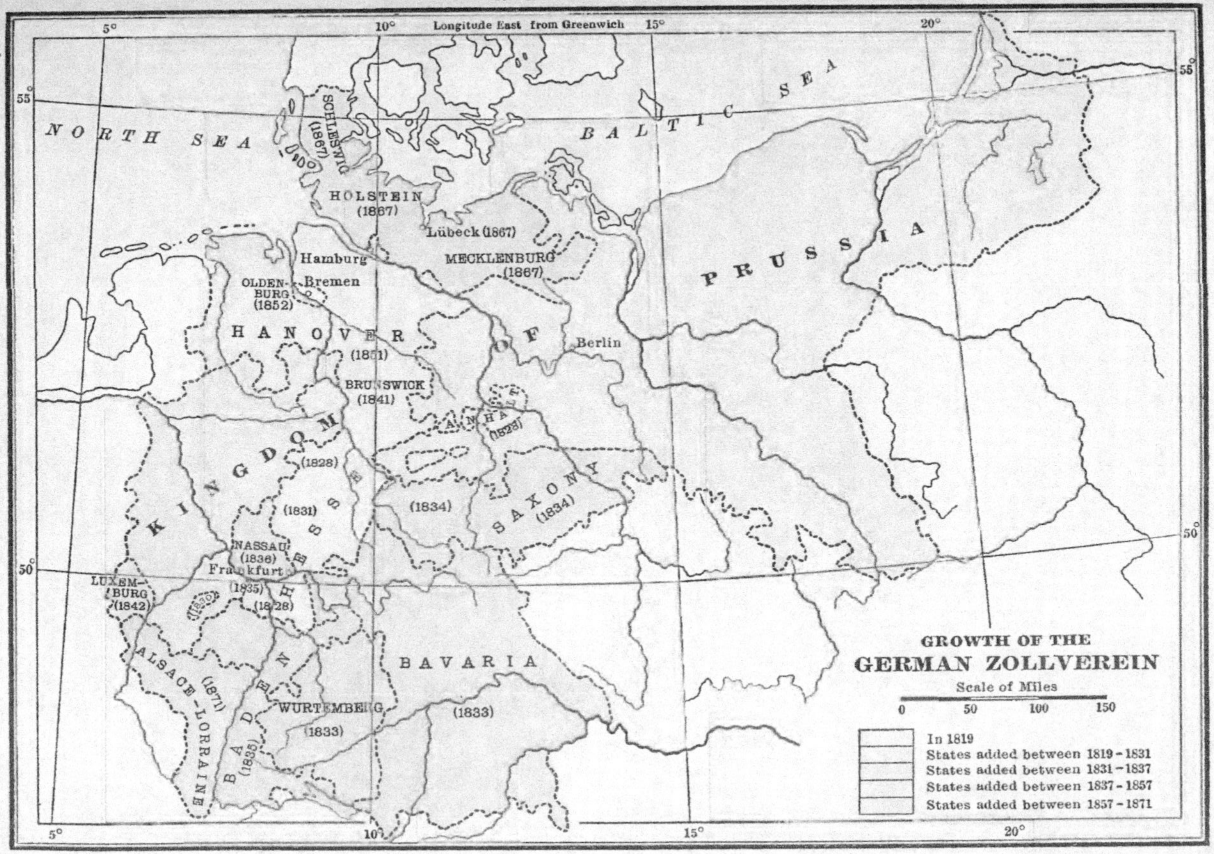
Longitude East from Greenwich
NORTH SEA
BALTIC SEA
SCHLESWIG (1867)
HOLSTEIN (1867)
Lübeck (1867)
MECKLENBURG (1867)
Hamburg
Bremen
OLDEN-BURG (1852)
HANOVER (1851)
BRUNSWICK (1841)
KINGDOM OF PRUSSIA
Berlin
ANHALT (1828)
(1828)
(1831)
HESSE
(1834)
SAXONY (1834)
NASSAU (1836)
Frankfurt
(1835)
(1828)
LUXEM-BURG (1842)
ALSACE-LORRAINE (1871)
BADEN (1835)
WURTEMBERG (1833)
BAVARIA
(1833)
GROWTH OF THE
GERMAN ZOLLVEREIN
Scale of Miles
0 50 100 150
In 1819
States added between 1819-1831
States added between 1831-1837
States added between 1837-1857
States added between 1857-1871

with the administration of these duchies in such a way as to embroil Prussia and Austria in 1866, and make the latter appear as aggressor. The consequences were the dissolution of the old confederation, the decisive Prussian victory of Sadowa or Königgrätz, the termination of the war in seven weeks, and Austria's exclusion from Germany. Prussia rounded out her territory by annexing small neighboring states which had sided with Austria, together with the two duchies which had been the bone of contention. She organized the North German Federation including all the territory of the present empire north of the River Main, in 1867. The South German states, Bavaria, Württemberg, Baden, and Hessen Darmstadt entered into a military alliance with Prussia, and it was evident that the slightest act of foreign aggression would create a flood of patriotism so irresistible as to carry away all the barriers between them. For this and other reasons Bismarck perceived the necessity of squaring accounts with the French Empire of Napoleon III.

A casual opportunity for discord was treated in such a way that France declared war, July 19, 1870. The South German states threw in their lot with Prussia, with whose system they had already brought their own military organizations into conformity. The military machine performed its function with bewildering accuracy and thoroughness. The military power of France was crushed in six months. By the Treaty of Frankfort, May 10, 1871, France was constrained to cede a territory of 5603 square miles, Alsace and a part of Lorraine, containing a population at that time of about 1,500,000, to pay an indemnity of five billion francs, and to grant to Germany in commercial relations the treatment of the most favored nations.

The successful war with France made a closer union of the states inevitable. The negotiations to this end were being carried on during the course of the war, and just before its termination, on Jan. 18, 1871, William I was proclaimed German Emperor in the Hall of Mirrors at Versailles in the presence of the princes of Germany and the generals of the army. Bismarck continued to guide the policy of his country as Imperial Chancellor until 1890.

A thorough application of the Principle of Nationalities in Europe would have removed most of the causes of international friction during the nineteenth century. It might very likely have eliminated the most immediate cause of the present war. Unfortunately many of the nationalities of Europe occupy territories so intricately arranged, and the geographical requirements of great political and commercial interests are so imperative, that it is doubtful whether the principle can ever be adopted in all instances. Some of the exceptions to the application of the doctrine have been a source of bitterness and suspicion, contributing in large measure to the irritation and inflammation of spirits which has developed concurrently with the more certain causes of the present war. It will contribute materially to our purpose to consider briefly the more noteworthy examples of the non-fulfilment or violation of the principles of national unity and independence in Europe.

Alsace and the part of Lorraine where French had not supplanted German as the spoken language were ceded by France to Germany in 1871. This territory was admitted by law to the German Empire, June 9, 1871, as the Reichsland, or Imperial Territory. The sovereignty is excercised by the Kaiser through his viceroy in Strassburg. On May 31, 1911 a new constitution with local self-government was inaugurated, the legislative organ being a diet of two chambers. The representatives in the lower chamber are elected by general, direct suffrage. Alsace and Lorraine have three votes in the Bundesrat, or Imperial Federal Council, and fifteen representatives in the Reichstag, or

Imperial Parliament. They contained a population of 1,874,014 in 1910, but their economic importance is out of all proportion to this number. Their mineral wealth was still largely a matter of conjecture at the termination of the Franco-German War. A subsequent development has shown that the richest iron deposits in Europe are concealed in an area embracing part of Lorraine, a section of northeastern France, and the Grand-duchy of Luxemburg. Thus the Reichsland makes an annual contribution of about 18,000,000 tons of iron ore to German industry, about four times as much as all Prussia. This fact is of capital importance in any considerations touching the disposition of the territory. France together with the land to which France lays claim is vastly richer in iron ore. But Prussia contains more abundant supplies of coal, and in the economy of industry the ore generally seeks the coal, not the coal the ore. Another very legitimate consideration of both a military and commercial nature must be duly weighed in connection with the question of the proper delimitation of the two countries, France and Germany. History teaches that rivers are not, as a rule, to be considered as suitable boundaries of national units. Mountains are far more adequate to provide the degree of isolation useful for setting off the national territory. Besides, rivers are highways of commerce, and therefore more appropriate as interior lines of communication with intimate political association of their two banks. For this reason, on geographic grounds, the line of the Vosges Mountains, and not that of the Rhine, is the appropriate boundary between France and Germany.

Union with any state must be treated as an exception to the Doctrine of Nationalities for the people of the Reichsland until the vexed question of their nationality is determined. They were part of the mediaeval empire, the Holy Roman Empire, which was established on a German foundation. Alsace became entirely French territory in 1697.

The historical situation respecting the present German Lorraine is more complicated. In the mediaeval period the term Lorraine was applied to a vastly larger territory than today. It even included part of the Low Countries. There was Upper and Lower Lorraine. Later the name was limited in its application to Upper Lorraine. The dukes of Lorraine in this restricted sense were in authority until well into the eighteenth century. Lorraine was nominally part of the Germanic body, but in reality stood in intimate relationship with France, with which it was finally united in 1766. France had annexed the Bishopric of Metz as early as 1552. The German Lorraine of the present time is a comparatively small part of the former duchy.

It appears that both geographically and historically Alsace is more appropriately attached to Germany than Lorraine; a circumstance which we might mildly put forward as possible basis of compromise in the famous controversy relative to the future of the Reichsland.

German blood predominates in the present Reichsland. But blood alone does not constitute nationality. Spiritual or cultural qualities outweigh it. For a long time the Alsatians did not assimilate with the French; but after the Revolution the upper classes at least became quite thoroughly Gallicized in spirit. The upper classes and clergy were very emphatic in their opposition to German influences after 1871. The people in the conquered provinces were required in 1872 to declare whether as individuals they chose to become German, or remain French citizens. At that time more than 150,000 announced their adherence to France, of whom nearly 50,000 removed across the border.

Conciliatory and harsh policies on the part of the German government have alike failed to bring about a complete assimilation of the people of the Reichsland. Many are

Helmuth Carl Bernhard, Count von Moltke, Chief of the General Staff of the army victorious over the French in 1870.

William I, King of Prussia and German Emperor.

Prince Otto Edward Leopold von Bismarck, creator of the German Empire.

reconciled to German rule, others are irreconcilables; but in neither case does this circumstance determine the question of nationality. The fact is, with the unique conditions of their historical experience, they have developed, like the Swiss, some independent characteristics of nationality,which tend to differentiate them in their sentiments and feelings from both the French and Germans.

Weighty problems of nationality await our attention in examining the situation in Russia.

In judging the charges against Russia for violating the Principle of Nationalities it will be well to adopt a broader point of view and consider Russia's territorial expansion in general, so as to survey the individual transgressions in their proper perspective. And at the outset, impartiality requires that we should not accept unchallenged the traditional impression of Russia's insatiable, inordinate, inhuman land-hunger, or land-avarice. It might seem to be a rather gratuitous undertaking to defend in this connection a state that has already absorbed one-seventh of the land surface of the world, and is apparently still striving after more. The following observations are only to suggest that the grounds for such a sweeping and unfavorable impression ought to be submitted to a rational examination, even though appearances seem so convicting.

Russia in Europe is a vast plain without internal natural barriers. The characteristic feature of Russian history has been an expansion over contiguous territory. We find no evidence in history to support a prior claim to the greater part of the Russian plain. The advance across Siberia began in 1581 and within a century the eastern boundary was carried to the Pacific Ocean. To condemn on ethical grounds the Russian occupation of Siberia would be as petty as to arraign the United States for extending its sway over the Great West disregarding the original sovereignty of the Red

Man. Peter the Great determined the principal aim of Russian policy regarding expansion, and to it Russia has steadfastly adhered, in her constant endeavor to acquire open ports. Peter founded his new capital in 1703, but its harbor is closed by ice four or five months every year. Archangel on the White Sea suffers from the same restriction. During the eighteenth century the present Baltic Provinces of Russia were annexed. Peter attempted to secure an opening onto the Black Sea. Catherine II secured the site and founded Odessa. The conquest of the southern seaboard from the Turks was really a recovery of old Russian territory. Russia took Finland from Sweden in accordance with an agreement with Napoleon I.

The immense empire of the Tsar constitutes a natural geographic unit. In general its boundaries are clearly defined. But on the south-west towards Austria-Hungary the political boundary had not, up to the beginning of the war, been carried forward to the Carpathian Mountains, which would constitute a suitable natural frontier. On the west the North German plain is like a boldly projecting angle of the Russian plain with no unmistakable intervening feature to define the limit of each. The political frontier between the Slav and Teuton powers has swayed back and forth throughout the centuries, and the fate of Poland, a Slav nation, is closely connected with the lack of natural defining features in this region.

The feeling of suspicion and aversion for Russia's imputed injustice in subjugating alien peoples rests mainly on conditions in Finland, the Baltic Provinces, and Poland.

Russia took Finland from Sweden in 1809. While the Finns are not a Slavic people, they are not Swedish either. There is no logical reason, therefore, why Finland should not continue under the protection of the Tsar, provided he respect the self-government and liberties of the people. The

Finns are a nation with historic personality. They are progressive and energetic. Their educational system has banished illiteracy from the land. Finland is a grand-duchy with the Tsar of Russia as grand-duke. The Tsars have sworn to uphold the fundamental laws, which cannot be altered without the consent of the diet. But an imperial manifesto in 1899 virtually abrogated the constitution by establishing that all general legislation, all legislation, that is, which had any bearing on the Empire as a whole, was henceforth to be enacted in the customary way, or in other words, in St. Petersburg. The Tsar was to decide what legislation was "general." At the same time the Finnish army was incorporated with the Russian.

A general strike in Finland at the time of the revolutionary disturbances in Russia secured a complete restoration of Finnish liberties and privileges in 1905. This was followed by a sweeping change in their constitution by the Finns themselves in 1906, when universal suffrage including both sexes was introduced. But the Russian government began again in 1908 to show a tendency to interfere in the self-government of Finland by demanding a closer control in financial and military matters, and at the present time the existence of the national independence of the Finns seems precarious.

The Baltic Provinces of Russia have no distinct nationality. The upper classes are largely German; the mass of the people Letts. Agitation among them has never had as its goal separation from the Russian Empire, or complete local independence. They wish to preserve intact their Protestant religion, the control of their schools, and the use of their language. But the situation is only in a limited sense a violation of the Principle of Nationalities.

The ancient Kingdom of Poland was destroyed by three successive partitions of her territory in 1772, 1792 and 1795, Russia, Prussia, and Austria participated in the first and last of these; Russia and Prussia acted without Austria in the second. In these partitions Russia received for the most part territory which was not essentially Polish, either land which had formerly been Russian, or regions where the population was mainly Lithuanian. Napoleon constituted the Grand-duchy of Warsaw, the nucleus of old Poland, at the expense of the Prussian and Austrian apportionments. The Congress of Vienna transformed this into the Kingdom of Poland and made Alexander I, Tsar of Russia, King of Poland, with the understanding that the connection with Russia should be merely personal, through the sovereign. Alexander I acting in good faith appointed his brother to rule Poland as viceroy, and granted the country a very liberal constitution. Mutual misunderstandings resulted in the failure of the arrangement. The Poles were discontented without the complete territorial restitution of their state. The Kingdom embraced only about one-sixth of the former territory. The unsuccessful revolution of 1830 led to the virtual annexation of the Kingdom of Poland to Russia. The Poles have never abandoned the aspirations for a restoration of their national existence. Now Russia stands pledged before Europe to preserve the independence of the territory entrusted by the Congress of Vienna to the sovereignty of Alexander I.

The verdict of the nations relative to Russia's attitude toward minor nationalities in future will be chiefly determined by her treatment of Finland and Poland.

The existence of the Austro-Hungarian monarchy, the union of the Hapsburg dominions seems to be a flagrant violation of the Principle of Nationalities. This political combination contains a population speaking no fewer than eleven distinct tongues, or nearer thirty if minor fragmentary languages are included. The nucleus of the political association is not constituted by a single prominent nationality as in the British Empire, but there are two principal centers, each represent-

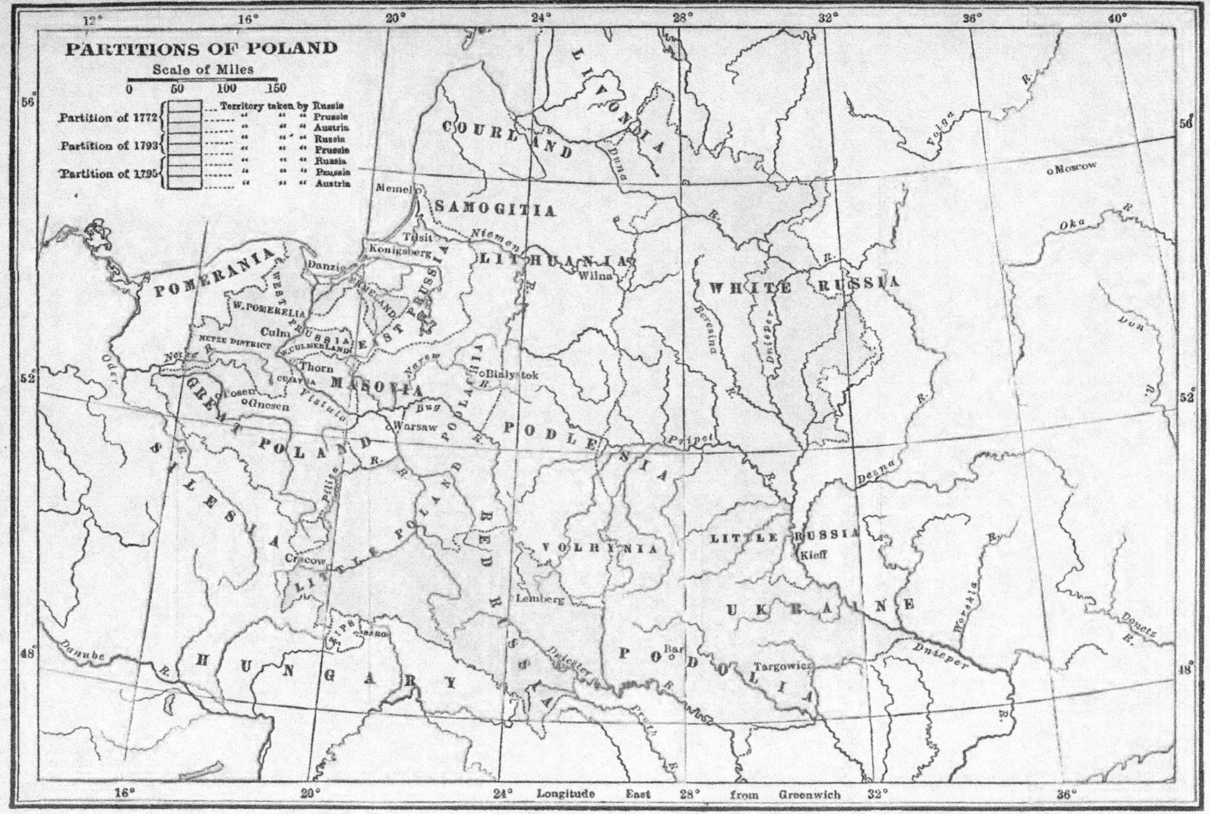
PARTITIONS OF POLAND
Scale of Miles
0 50 100 150
Territory taken by Russia
Prussia
Austria
Partition of 1772
Partition of 1793
Partition of 1795
Longitude East from Greenwich
COURLAND
LIVONIA
SAMOGITIA
LITHUANIA
WHITE RUSSIA
POMERANIA
WEST PRUSSIA
W. POMERELIA
NETZE DISTRICT
MASOVIA
GREAT POLAND
SILESIA
PODLACHIA
PODLESIA
VOLHYNIA
LITTLE RUSSIA
LITTLE POLAND
RED RUSSIA
UKRAINE
PODOLIA
HUNGARY
Memel
Tilsit
Konigsberg
Danzig
Culm
Thorn
Posen
Gnesen
Warsaw
Bialystok
Wilna
Cracow
Lemberg
Kieff
Bar
Targowica
Moscow
Niemen
Duna
Beresina
Dnieper
Pripet
Desna
Vistula
Narew
Bug
Oder
Netze
Pilica
Dniester
Pruth
Danube
Volga
Oka
Don
Worskla
Donetz

ing one of the two major partners in the society, the Germans and the Magyars, or Hungarians. The agglomeration of lands and heterogeneous peoples under the rule of the Hapsburg family was due in large measure to a succession of fortunate matrimonial alliances. An old proverb celebrated the good-fortune of the Hapsburgs, who could increase their estates by this amiable method: *Sed tu, felix Austria, nube.* Their possessions were formerly more scattered; but the far-reaching territorial pruning processes of the past hundred years or more have lopped off the too diffuse outgrowths of Austrian sovereignty. The Austrian Netherlands were absorbed in the first French Republic, and included in the Kingdom of the Netherlands in 1815. The Hapsburgs were shorn of their Italian possessions in the manner which we have observed. But what the Austrian domains have lost in extension has been more than counterbalanced by the greater advantage of territorial compactness. To-day the monarchy forms in general a convenient geographic unit, although its surface presents the most varied features.

Hapsburg policy depended largely on the ability to establish a sort of equilibrium of racial jealousies, by opposing the interests of one nationality to those of another. Emperor Francis I is reported to have formulated the guiding principles of his statecraft as follows:

"My peoples are strangers one to another. So much the better. When fever attacks the body politic in France, it seizes you all the same day. I send Hungarians to Italy, and Italians to Hungary. Each watches his neighbor. Their incapacity to understand one another gives rise to mutual hatred. Public order is based on their antipathy, and their enmity is the guaranty of general peace." This attitude of suspicion is exhibited by the maintenance of the internal customs barriers separating provinces much longer than in Prussia. Popular institutions develop early where there is national homogeneity as in England. They were naturally very tardy in their development with the conditions prevailing in the Austrian dominions. National unity is almost indispensable to a sound popular form of government.

History alone can explain the causes of the peculiar institutions of the Austro-Hungarian Monarchy. Why, for instance, have the Hapsburg dominions, with their varied population, a dual rather than a federal constitution? Dualism is the outcome of a long contest. Its chief cause is the tenacity of the Magyars in adherence to a national ideal. Hungary was crushed in a military sense in 1849. Her constitution was abrogated, and she was treated as conquered territory; but her spirit did not yield. Her people adopted an attitude of passive protest. In 1861 the Austrian government adopted a conciliatory policy. Francis Joseph promulgated a liberal, federal constitution for all his dominions. The representation in the popular house of the imperial parliament was to be apportioned among the nationalities according to population.

It often happens that one of the partners in an association by sheer stubbornness, by a determination to sacrifice every interest of his associates and even himself rather than fail to assert his will, makes himself the dominant factor. The others prove eventually to be more compliant and go to him because he will not go to them. This was the rôle of Hungary. The imperial parliament assembled, but no Hungarian deputies appeared, and from 1861 until 1867 the Magyars persisted in their invincible campaign of negative resistance. By no official act would they recognize the existence of the new institutions. Their objections were varied. In the first place Hungary was a historic kingdom with definite boundaries and its own traditions and ideals. The Magyars felt that they would be untrue to their historic personality if they accepted such a collective arrangement. The head of the House of Hapsburg could claim their lawful allegiance solely as King of Hungary, after he had taken the oath to uphold their fundamental laws. Any proposals touching their constitution or diminishing their independence were a subject for deliberation between themselves and their monarch, and no change could be made without their formal sanction. It was only a casual circumstance to them that their king happened to be ruler at the same time of other lands and peoples. The plan for a federal union was distasteful for other reasons as well. The Magyars were out-numbered in their own territory by other nationalities, Roumanians and Slavs. The Magyars claimed the right to hold these nationalities in political subordination, and their pride revolted at the humiliating prospect of a union on terms of equality with their subjects.

The situation was like Mohammed and the Mountain. Francis Joseph finally resolved to go to the Mountain. The military disaster in 1866 and the exclusion from Germany were persuasive factors in this decision. The Ausgleich, or Compromise, of 1867, was the consequence. It has remained the organic law of the Austro-Hungarian Monarchy.

It has been expedient to unfold the persistence of Hungarian temperament and its result at greater length, because it remains the most distinctive quality in the political nature of the Monarchy to this day.

The Compromise recognized as a fundamental principle the independent existence of the two states, Hungary, including Transylvania and Croatia, and the Austrian Empire, comprising all the other Hapsburg possessions. The two states have the same sovereign, who is king in Hungary, emperor in Austria. They form an association with joint ministries for a limited range of common interests, foreign affairs, war, and finance. Each parliament choses a delegation of sixty of its members to supervise the three joint ministries. These delegations meet alternately in Vienna and Buda-Pesth. There is a common tariff system, and the proceeds of the customs are applied to the common expenses, the expenditure in excess of the income from customs being contributed by the states in the proportion of 63.6% for Austria and 36.4% for Hungary. The joint expenses amount to about $120,000,000 annually.

The Compromise was satisfactory to the Germans and Magyars only; for they are the dominant elements, although each people constitutes a minority in the state in which it occupies the position of political prominence. The other races refuse to acquiesce in the dual arrangement as final. Even the relations between Austria and Hungary, particularly at the periodic regulation of common financial affairs, have been the source of so much friction, that if Francis Joseph had scrupulously endeavored to impersonate consistently the double rôle of king and emperor, he must frequently have found himself in the embarrassing situation of the Lord Chancellor in Gilbert and Sullivan's "Iolanthe".

The policy of Austria has been more liberal than that of Hungary. The Austrian Empire as partner in the Dual Monarchy is divided into seventeen provinces, each with its own diet for local affairs. The Austrian ministry is responsible to parliament. Universal suffrage was granted in 1907, and in the first elections on the broader basis the Socialists polled nearly one-third the votes.

The Kingdom of Hungary includes Transylvania and Croatia. The Hungarian policy has been characterized by a determined effort to Magyarize the subordinate nationalities, as shown in the refusal to allow them the use of their own languages in official business. Croatia has in large measure been excluded from the scope of this process. The franchise is very restricted in Hungary, because the proportion of non-Magyars is much higher among the poorer classes.

Bosnia and Herzegovina have been administered conjointly by the two states.

The present size and population of the Monarchy and its fundamental parts are as follows:

	Area in Sq. Miles	Population
Austrian Empire	115,831	28,571,934
Hungary	125,641	20,886,387
Bosnia and Herzegovina	19,767	1,931,802
Total	261,239	51,390,123

In the Austrian Empire, German is spoken by 36% of the population; Bohemian, Moravian, and Slovak by 23%; Polish by 17%. In Hungary, Magyar is spoken by about 48%; Roumanian by 14%; the other languages are all members of the Slav family. If we consider the Monarchy as a whole, the Germans (12,010,669) and Magyars (10,067,992) form together (22,078,661) considerably less than half the entire population. They are outnumbered together by the total of the Slav peoples; and yet the Slav element is disregarded in the organic partnership upon which the Monarchy is based.

Before dismissing the subject under discussion we must consider briefly the fragments of the Polish nation which are not included in the Russian Empire. After the settlement of European boundaries in 1815, Galicia was retained by Austria, and West Prussia and Posen by the Kingdom of Prussia, all three having been parts of dismembered Poland. The treatment of Poles in the Austrian and Prussian territories presents a striking contrast. In the former they enjoy autonomy, and even an opportunity of oppressing others, in the latter every means is employed to crush their national spirit.

Galicia occupies the north-eastern part of the Dual Monarchy between the Carpathian Mountains and the Russian frontier. It embraces about 30,000 square miles and contains a population of about 7,000,000. Geographical conditions associate it with Russia, as we have already observed, but a majority of the inhabitants would probably not welcome with much enthusiasm a political union with the Russian Empire, although they are nearly all Slavs. For Galicia is attached to the Austrian division of the Dual Monarchy, and the policy of Austria in respect to the subordinate nationalities is much more tolerant than that of Hungary. Thus in the Austrian Empire, as distinguished from the Kingdom of Hungary, there are seventeen diets, legislating in local affairs; in Hungary there is only one such subordinate legislature, the diet of Croatia. Galicia enjoys a larger measure of autonomy than any other part of Austria. But Galicia is divided between the Poles in the west and the Ruthenians in the east, both of them branches of the Slav family, but the latter more closely related to the Russians. The Poles are Roman Catholics; the Ruthenians are partly Uniates, partly members of the Russian Orthodox Church. The Uniates retain the Slav liturgy but acknowledge the supremacy of the pope. The distinction between the Poles and Ruthenians has caused much friction in Galicia. Austrian policy has favored the Poles, and has refused independent autonomy to the Ruthenians, thus subjecting them to the Poles who are somewhat more numerous, as well as superior in culture and wealth.

The Kingdom of Prussia had an awkward outline before it acquired its present Polish possessions. For West Prussia, which was Poland's only outlet to the Baltic Sea, divided the

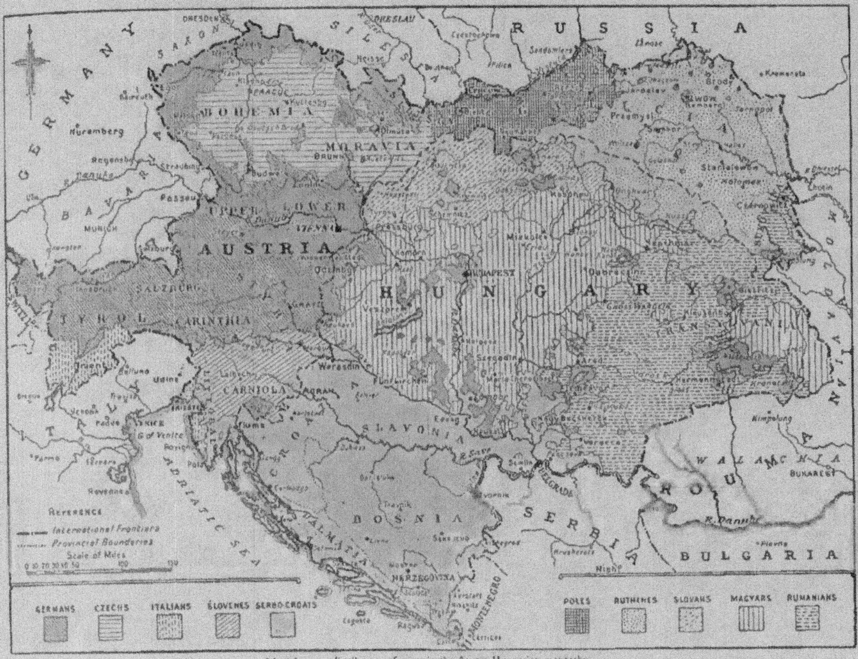

Map showing distribution of races in the Austro-Hungarian monarchy.

Hohenzollern realm in two, cutting off East Prussia from the main part of the territories. Fortunately for the compactness of Prussia, Germans are now in the majority in West Prussia.

The case is far different in Posen. This province contains about 11,000 square miles with a population of two million. In outline it is like a wedge driven into the eastern flank of Prussia. As a part of Prussia it makes the sharp angle between the outlying eastern stretches of Prussian territory less acute. The Polish population is all Roman Catholic. Not only are the Poles a very large majority in the population of Posen, but their number tends to increase more rapidly than that of the Germans by reason of their larger birth-rate, and a constant immigration from Russian Poland. They are also advancing rapidly in prosperity. Prussia has regarded this development of Polish nationality with intolerance. Partly in view of the great strategic importance of the province of Posen, the Prussian government inaugurated a vigorous process of Germanization in the Polish districts, which has been largely a failure. In the first place, they insisted that the German language should be the only medium of instruction in the schools wherever possible, and instructed the police to break up public meetings conducted in Polish. Then, in 1888, they adopted the policy of substituting German for Polish farmers, established a commission for buying lands of the Polish owners so as to rent them in turn to German colonists, and devoted nearly $25,000,000 to this purpose. A like sum was added to this ten years later.

History affords ample proof that persecution often stimulates the cause which it is intended to stifle. Even in Prussia, the government's policy was severely criticised, when it was learned, in 1902, that Polish children had been subjected to corporal punishment for refusing to repeat the

Lord's prayer in German in the schools, while in some instances the parents, who had interfered, had been fined or imprisoned. Count von Bülow, as president of the Prussian ministry, admitted in parliament that the process of Germanization had thus far been a failure, but he persuaded them in 1903 to place about $62,500,000 at the disposal of the commission for purchasing Polish lands which were to be sold to German colonists.

The repressive policy of Prussia culminated in a bill for the compulsory expropriation of Polish lands in Posen and West Prussia, which was introduced in the Prussian parliament by Prince von Bülow, in 1907, and became a law the following year. As amended, the appropriation for land purchases contained in the bill amounted to about $68,750,000. Prince von Bülow asserted that the Poles had not appreciated the advantages of Prussian rule, but adhered stubbornly to their own nationality. This law was denounced by an important part of the press and public opinion. The circumstances tempt us to make a comparison with the agrarian policy of the British government in Ireland. In the Polish provinces, the Prussian government had voted sums aggregating about $181,000,000 to secure the eviction of Poles, the subordinate race. In Ireland, the British government down to 1906 had lent its credit for more than $155,000,000 to facilitate the purchase of land by the conquered, and the spoliation of land-lords, who as a class represented the conquerors.

At the same time as the Prussian land bill, the Reichstag passed a law regarding public meetings, which prohibited the use of languages other than German with certain exceptions, as during the period of political campaigns, and in international congresses. In their attitude towards languages the Germans present this striking paradox; as individuals, they aim to make themselves polyglots, and

The monument to Prince Bismarck in Hamburg, dedicated June 2, 1906; erected by public subscription from designs by Lederer and Schaudt.

Monument at Leipzig, commemorating the battle of Leipzig, in 1813, when the Austrians, Russians, Swedes, English, and Prussians defeated the French under Napoleon I.

their plan of education is wonderfully adapted to fulfilling their aspiration for proficiency in foreign languages; but as a nation, they sternly repress any alien tongues that linger in sections of their territory. A single example will serve to show to what extremes this jealous attitude toward alien tongues can be carried. In North Schleswig, where the majority of the people are Danish, and therefore suspected of separatist tendencies, the authorities refused to permit Captain Amundsen, the Norwegian explorer, to deliver a lecture in his own language, which differs only slightly from Danish, and would, therefore, be understood by the local population. Later the Prussian Minister of the Interior withdrew the prohibition.

We must now turn our attention to the other great principle, which competes with the claims of nationality for the position of supremacy in the preoccupations of statesmen.

The endeavor to maintain a balance of power, or equilibrium of forces, between the more important states has been a dominating factor in European diplomacy, certainly since the time of Louis XIV. Great Britain is accused with the greatest bitterness of the policy of fomenting antagonism among the continental states for selfish aims in upholding this principle against whatever state was most powerful. However, the states which joined Great Britain in coalitions with a view to restoring the equilibrium are equally culpable, if the undertaking was reprehensible, and some of those who condemn this conduct most emphatically on other occasions uphold the theory that political egoism is the appropriate guiding motive for the foreign policy of a state. Moreover, Great Britain has only intervened once against a continental power since the down-fall of Napoleon, and in that single instance the general principle of a European equilibrium was not the principle motive. It seems petty, therefore, to employ this subject as a pretext for international vituperation.

The idea of even an approximately exact balance of power is absurd. It would be conceivable only if national areas could be adjusted under conditions of such ideal uniformity as prevailed on our western prairies before the period of settlement. If applied logically in Europe the principle of equilibrium would mean the violation of the Principle of Nationalities, because the conditions controlling the development and power of states are naturally of infinite variety. They cannot be reduced to uniformity. To overcome this difficulty, diplomacy has devised enduring alliances, groups of states of different sizes, so that the aggregate strength of such associations may be equivalent, just as a mason builds up walls of equal height, even though no two stones in them are of the same size. Unfortunately the units with which European diplomacy deals are too few in number to permit of offsetting entirely the individual discrepancies by combinations. It cannot be maintained, however, that the result of this policy has been entirely a failure, nor that the alliances have been entirely without influence in preserving the peace.

The two alliances with which we have to deal are the Triple Alliance between Germany, Austria, and Italy, and the Dual Alliance uniting France and Russia.

The maintenance of friendly relations with Russia was the traditional policy of the House of Hohenzollern. The mutual attitude of the two governments was very cordial throughout the greater part of the reigns of William I of Prussia and his nephew Alexander II of Russia. Russia declared her readiness in 1870 to intervene against Austria if she should stir to help France. After the termination of the war there were increasing signs of a friendly spirit in the relations of Germany and Austria. An interview of the three emperors and their ministers was arranged at Berlin, Sept. 1872, and the consequence of a neighborly interchange of views was an understanding, little less effective than an alliance, which endured for about six years. The interviews became annual affairs and the visits of the King of Italy at Vienna and Berlin in 1873 opened a prospect of a further extension of the association. There were no inherent causes of discord in the relations between Germany and Russia, but unfortunately the policies of Austria and Russia were of such a nature that misunderstandings would almost certainly arise; for both pretended to have vital interests in the Balkan peninsula. The time came when Germany had to choose between her two friends, and preference for one would almost inevitably alienate the attitude of the other. The cold impartiality displayed by Bismarck in directing the proceedings in the Congress of Berlin wounded the feelings of the Russian representatives. Bismarck had made his choice and committed Germany to it by a formal treaty of alliance for peace and mutual defence with Austria in October, 1879. According to the terms of this alliance, as made public in 1888, the powers pledged mutual assistance in case either were attacked by Russia; they promised friendly neutrality in the event that either were attacked by any other power. Italy, resenting the annexation of Tunis by France in 1881, joined the alliance in 1883, thus expanding it to a triple alliance. The terms of agreement were undoubtedly modified and extended to bring them into harmony with the new conditions and broader scope.

France had been isolated since the war. Bismarck had repeatedly declared that Germany intended to maintain the peace of Europe. But the rapidity of her successes and the incontestable superiority of her military establishment kept alive a feeling of mistrust. The bill for the reorganization of the French army in 1875 was the occasion for a protest by the German ambassador in Paris and a demand by some German newspapers that France be crushed more thoroughly before she should have time to recover. The danger of German aggression is said to have been averted in 1875 by a friendly intervention of Russia and Great Britain at Berlin. But France did not enjoy a feeling of assurance as long as her isolation continued. The impulse to closer relations between France and Russia seems to have originated in the latter country. The friendship between Russia and Germany had apparently not been destroyed by the important event of 1879. Alexander III, although in general hostile to German influence, was determined above all to maintain peace. Accordingly, at an interview of the three emperors at Skiernevice in September, 1884, an agreement was solemnized between Russia and Germany, each binding herself to maintain friendly neutrality in case the other were attacked by another power. This agreement was commonly called the Re-insurance Treaty.

The German government refused to renew this agreement in 1890. In the meantime the possibility of an alliance between Russia and France had been discussed from time to time in the former country since 1879. The motives of Russia in drawing near to France were not solely political. Russia was in need of capital. She stood at the threshold of her era of industrial development. Count Witte, the guardian angel of this movement, as Minister of Commerce, urged the advantage of attracting foreign capital to Russia. Paris was the best available money market. A Russian loan had been negotiated there in 1891. Probably from that time an understanding existed between the two countries. The interchange of the customary international demonstrations of cordiality indicated the development of the understanding into the formal Dual Alliance, the existence of which the visit of Tsar Nicholas II to Paris in 1896 was calculated to confirm.

The palace at Peterhof on the Gulf of Finland. Built by Peter the Great in 1720.

The Palace of Sans Souci at Potsdam, the "German Versailles." Built by Frederick the Great in the eighteenth century.

CHAPTER II

The Nations on the Threshold of the Twentieth Century

Great Britain: industrial and commercial supremacy in the nineteenth century; democratic development; imperialism and social reform; Egypt; South Africa; Imperial Federation Movement and Trade Preference; Irish Question; social legislation; Lloyd George's famous budget; suppression of Lords' veto; Home Rule Bill. France: miraculous salvation of the Republic; Boulanger; Dreyfus Case; Radical *bloc*. Russia: reforms and revolution; constitutional régime; peasant problem, agricultural Russia; Industrial Revolution in Russia, present condition of industry; commerce; government finances; German opinion of the condition of Russia. Germany: Prussianization of Germany; the army as a school of citizenship; the imperial and Prussian governments, measure of popular influence, political parties, Center, Conservatives, progress of agriculture, National Liberals, Social Democrats, Socialist program, state socialism; progress of German industry; government finances; general comparisons of Germany, United Kingdom, and France; the world's iron ore supply.

Great Britain occupied an enviable position among the states of Europe during the greater part of the nineteenth century. The industrial revolution, which had been inaugurated there in the second half of the previous century, did not affect the life of the continent until after 1815, did not triumph in Germany until after 1850, and hardly appeared in Russia before 1890. It placed England fifty years ahead of her nearest competitors. The great naval victories of the Napoleonic War and the driving of her rivals from the seas consolidated her maritime preëminence. Thus the industrial and commercial supremacy of the United Kingdom was firmly established in the early years of the century. It was never seriously challenged until after 1870. In some respects it has never been overthrown. The ascendancy of the British Islands in the textile industries and the world's carrying trade is still maintained. (See the table of statistics, page 79, Nos. 12 and 22). Englishmen formed the habit of assuming an unquestioned superiority in industry and trade, because their position had seemed unassailable for so long a time. Thanks largely to the profits of trade and the enterprise of her people as displayed in world-wide activity, England came to possess in London the leading money market in the world.

It is remarkable that Great Britain attained so lofty a position in material prosperity with a government which was not only unbelievably antiquated in its forms, but which was abounding in injustice, abuse, and corruption. And it is a greater wonder that at the time when the British parliament was an object of high regard on the continent, when it was making its great contribution to the political institutions of other nations by serving as a model of supposed excellence, parliamentary institutions in Great Britain were appallingly unreasonable and debased.

Parliament was the real ruler, and the real executive was the ministry of the hour, which was dependent for its tenure of office upon the majority in the House of Commons. But the House had become the organ of a small governing class in consequence of the archaic regulations for the election of its members. The apportionment of representation was glaringly unjust, and the suffrage was very restricted. Some of the boroughs retained their parliamentary privileges although they had gradually lost all their inhabitants, so that it was said that three niches in a certain stone wall sent two representatives to parliament, and a certain park, where no houses were to be seen, returned two others. Bribery, moreover, was carried on openly and on a vast scale. The great mass of the British nation did not acquire the opportunity for political articulation until late in the nineteenth century.

The political enfranchisement of the masses was obtained by three successive Reform Acts which broadened the basis of suffrage and redistributed the seats in the House of Commons. The bill of 1832 gave the franchise to the middle class, the act of 1867 extended it to the better class of laborers in the towns, and the final measure of 1884 admitted the majority of the laboring class, urban and rural. In the early part of the century three-fourths of the children received no instruction. There was no national system of schools until 1870. The chief interest in recent English history attaches to the fact that it deals with the period when the mass of the nation is finally master of its destiny and has received sufficient education to consider rationally what it wants. Two political tendencies have been clearly distinguished in the later period, the tendency to emphasize the importance of the empire, and the tendency to undertake great social and economic improvements at home. Toward the close of the period, with the ever increasing political intelligence of the people, the interest in social legislation seemed to be gaining decidedly the upper hand.

Two administrations, standing at the threshold of the age which we are considering, strike in turn the key-notes of the two motifs which are intermingled in the subsequent composition without losing their identity. The Gladstone administration of 1867–1874 devoted its attention almost exclusively to domestic reforms, carrying through Irish Disestablishment, the Irish Land Act of 1870, and the Education Act of 1870. The Conservative administration of 1874-80 turned its face to external affairs. The Prime Minister, Benjamin Disraeli, later Lord Beaconsfield, purchased for the nation the Suez Canal shares of the Khedive in 1875, undertook to maintain a vigorous attitude in connection with the Eastern Question, and annexed the

Transvaal, or South African Republic, in 1877. The Prime Minister posed as the advocate of the imperial idea. It was the beginning of modern British imperialism. As early as 1872 he had said:

"In my judgment no minister in this country will do his duty who neglects any opportunity of reconstructing as much as possible our colonial empire, and of responding to those distant sympathies which may become the source of incalculable strength and happiness to this land."

His versatile imagination conceived an appropriate symbol for his romantic imperial conception. On January 1, 1877, Queen Victoria assumed the title Empress of India.

The foreign problems of this ministry outlived it. In 1882 the Gladstone ministry which succeeded it had to intervene in Egypt to protect British interests; and the consequence of this was the British occupation of the country which has continued until the present. The task of reconstruction was entrusted to Lord Cromer as British Consul-General in Egypt. The Boers rose in revolt and defeated a small detachment of British troops at Majuba Hill, Feb. 27, 1881. Gladstone, who had already been occupied with the project of restoring the independence of the Transvaal, had the courage to persevere in his conviction of justice, although he was severely criticized at home, and the Boers accepted his concession as the prize of victory. By the Pretoria Convention in 1881 their independence was recognized subject to British suzerainty. This arrangement was modified by the London Convention in 1884, when mention of suzerainty was omitted, and the name South African Republic was restored, Great Britain retaining control of the foreign relations.

This same year gold was discovered in great quantities in the Rand, which attracted a large number of outsiders, Uitlanders, as the Boers called them, and created a very

Statue of Frederick the Great in Berlin.

Statue of the Tsar Alexander III in Moscow.

serious problem. The Boers, who were exclusively engaged in farming and grazing, and were the embodiment of simplicity in their habits, were menaced with submersion by the flood of immigration. They refused to grant the rights of citizenship on reasonable terms, and yet laid the chief burden of taxation upon the Uitlanders, who were henceforth, in a material sense, the progressive element in the population. The resulting complaints led to the famous raid, or filibustering expedition of Dr. Jameson, administrator of Rhodesia, in 1895. Those who designed this undertaking did not appreciate the capacity for resistance of the Boers. Dr. Jameson and his followers were quickly made prisoners and turned over to the British authorities, tried in London, and given light penalties, a fact which did not tend to appease the suspicion of the republican burghers. These were the days when the tide of aggressive imperialism was running high in Great Britain, when a spirit like that of some of the conspicuous German jingoes of to-day animated a large portion of the British public. On the other hand, there seemed to be reason for believing that the Boers of the Transvaal were carrying on a propaganda in the Cape Colony, where the Dutch outnumbered the British, with a view to creating a united South Africa with the Dutch race in supremacy. The Transvaal government seemed to be playing the rôle which at present Austria-Hungary imputes to Serbia. War was practically inevitable from the time of the Jameson raid. It broke out in October, 1899, and continued nearly three years. The Orange River Free State threw in its lot with the South African Republic. The subjugation of the Boers required an enormous expenditure, and the presence of 250,000 men constantly in the field. As soon thereafter as it was reasonably possible, the British government (a Liberal cabinet had succeeded to the Conservative war-ministry in 1905) granted self-government to the Transvaal (1906) and the Orange River Colony (1907), and the two provinces united with Natal and Cape Colony to constitute the South African Union, in 1910. Causes of discord in South Africa since the Union have been Premier Louis Botha's enthusiasm for imperial defense, problems arising from the presence of large numbers of indentured East Indian laborers, and serious strikes in 1913-14.

An increasing pride in the extent and resources of the British Empire, and the prominence of colonial affairs have fostered a desire for a closer political union. This aspiration has found expression in a series of colonial conferences in London, beginning in 1887, on the occasion of the fiftieth anniversary of Queen Victoria's accession to the throne. After others had been held in 1897 and 1902, the conference in 1907 established the rule that they should assemble at regular intervals, every four years. Thenceforward they have been known as Imperial Conferences. Although the British Empire contains more than 420,000,000 inhabitants, not more than 60,000,000 belong to the British race. Any schemes for imperial federation are limited to the latter. In 1911, Sir Joseph Ward, Premier of New Zealand, proposed an imperial parliament of two houses; but the other representatives were not prepared to go so far. The difficulties are great,—to define its powers, to distribute representation, and to determine its relationship with the legislatures now existing, notably the British Parliament. A more fruitful field of endeavor has been opened by the proposal to form a commercial or customs union. The chief difficulty in this connection has been the traditional free trade policy of Great Britain. Mr Joseph Chamberlain, the enthusiastic imperialist, favored the introduction of protective duties in the United Kingdom, so that while the bars were let down all around within the Empire, there would still be everywhere a defensive tariff barrier against the world outside. But the overwhelming defeat of the Conservatives in 1906 seemed to be a popular verdict against the proposed higher tariff in Great Britain. Canada and Australia both grant a preferential treatment to imports from the other parts of the empire. Another practical imperialistic question has been that of the colonies assuming a share of the burden of common defense. Australia had already contributed a warship, when, in the summer of 1912, Mr. Borden, Prime Minister of Canada, and some of the other Canadian ministers consulted informally with the imperial authorities concerning a much more extensive project for Canada. During the next session a bill passed the lower house of the Dominion Parliament to appropriate $35,000,000 for the construction of three super-dreadnoughts as Canada's contribution to imperial naval defense. The bill was rejected by the Senate, May 13, 1913, until the people could express their judgment of the matter in the next general election.

Undoubtedly the attempts to bring about a closer association of the scattered dominions have been watched with lively interest, if not apprehension, by other nations. The proposal to modify the fiscal policy was a matter of grave concern to Germany; for German trade had enjoyed the advantage of the "open door" throughout the British Empire, which was Germany's best customer. Moreover, naval assistance from the colonies might soon relieve Great Britain from the necessity of sending any considerable part of her own ships to distant stations. In Germany, where the national history had developed exclusively on a continuous territorial area, and where experience had emphasized the need of force in rendering effective the consciousness of unity, the dispersed members of the British Empire, which it would be impractical to coerce, seemed to be a very insubstantial material for the erection of a strong, united nation. It was predicted that the colonies awaited only an opportunity to go their own way. The rejection of the naval proposal by the Canadian Senate, even though it had been adopted by the popular house, was received as corroborative evidence of the opinion already held; the colonies were unwilling to concern themselves with Britain's problems.

The recent domestic politics of Great Britain far outweigh in fundamental importance, and even in sensational interest, the course of imperial affairs. A large part of Gladstone's career was spent in endeavoring to solve the Irish Question. During his third administration he introduced the first Home Rule Bill, April 8th, 1886, which led to the disruption of the Liberal Party, the Liberal Unionists withdrawing from it, and resulted in the defeat of the ministry. Lord Salisbury succeeded Gladstone with a Conservative cabinet, which secured the passage of an act to facilitate the purchase of their holdings by the Irish tenants. The Conservatives' policy, as later expressed, was to kill Home Rule by kindness. Their land-purchasing policy has since been greatly extended by an act in 1903, so that to-day the condition of the Irish peasantry has been economically vastly improved. During the next Liberal administration (1892–1895) Gladstone brought in a second Home Rule Bill, which was passed by the House of Commons, but thrown out by the Lords. The next Conservative administration (1895–1905), first under Lord Salisbury, then Mr Arthur Balfour was dominated, as we have seen, by imperialistic interests. The Radicals have been in power since 1905, and their legislation has aimed at an extensive transformation of established usages. They have been fearless, impatient opponents of deep-rooted social injustice. They have introduced startling innovations into British institutions. Their opponents characterized their policy as revolutionary; but a thoughtful consideration of the record of the past ten years reveals the temper of the great majority of the nation. They had roused themselves to carry out a thorough renovation of the whole political structure, and all considerations of foreign policy were subordinate to this purpose. Sir Henry Campbell-Bannerman was Prime Minister until April 5, 1908, when he was succeeded by Mr. Herbert Asquith.

The first great reform was the introduction of old-age pensions, taking effect Jan. 1, 1909, and adding more than $35,000,000 to the budget. It was looked upon as justice, not charity. This measure and the expansion of naval expenditure increased the requirements of the government by more than $80,000,000. Mr. Lloyd George, Chancellor of the Exchequer, met the unprecedented situation by a famous budget, by the provisions of which he proposed to promote certain social and economic reforms as a by-product in the process of taxation. He laid a tax on unused land, and appropriated for the public a portion of the unearned increment in land values. This budget was rejected by the House of Lords, Nov. 30, 1909, which resulted in a general election, in which the abolition of the Lords' veto and Home Rule were essential parts of the Radical program. The resulting strength of parties in the House of Commons, not sensibly altered since, was as follows; Unionists (Conservatives), 273; Liberals, 275; Laborites, 40; Nationalists, 71; Independent Nationalists, 11. The Parliament Bill was introduced, May 10, 1910. It provided that the Lords should have no power to veto a money bill, and that any other bill passing the Commons in three successive sessions could be submitted to the king for his approval in spite of the veto of the Lords, if two years had elapsed since it was first introduced. After a spirited contest, during the course of which there had been a fresh appeal to the country without material change in the numerical relation of parties, this bill was accepted by the Lords, Aug. 10, 1911. It cleared the way for the Irish Home Rule Bill, which was laid before Parliament, April 11, 1912. A threatening agitation began to spread through the Protestant population of Ulster, where the prospect of association with the rest of Ireland in autonomy was viewed as a threatening calamity. A covenant of resistance was sworn to, Sept. 18, and Ulster volunteers were enrolled with Sir Edward Carson as the animating spirit in the movement. The Ulsterites threatened to set up an independent provincial government as soon as an Irish parliament should meet at Dublin. The Home Rule Bill passed the House of Commons for the third time, May 25, 1914, in spite of the demand of the opposition for a referendum. As it had been rejected twice by the Lords it eventually received the royal assent. The situation became continually more tense, and civil war appeared to be imminent. It is obvious that the nation was absorbed in these internal contentions, and was comparatively indifferent to foreign issues in the first half of 1914. The period of aggressive imperialism had passed, and the zeal for social reform which animated the government was proof of a sincere desire for the preservation of peace. We shall discover that France, too, had been largely engrossed in remodeling her institutions.

The French Republic, which had been proclaimed in Paris, Sept. 4, 1870, upon reception of the news of Sedan, was accepted without enthusiasm by the majority of Frenchmen at the termination of the war. Some looked upon it as a merely temporary expedient, others as a permanent necessity. It had proved a failure before, but so had all other forms of government, and therefore the obvious thing to do was to accept what circumstances had thrust upon them. It was a period of weariness, disillusionment, and exhaustion. The greater glory is due to a few souls of loftier, far-reaching vision, who combined unusual practical ability in statesmanship with irresistible enthusiasm for an ideal. In consequence of their devoted efforts, the state was quickly reorganized, and the Republic began to acquire the devotion of the citizens.

A unity of purpose in the apparently confused course of French history in the 19th century is perceived, if we consider the whole period since 1789 as the revolution, a revolution of which the different outbreaks were single incidents, and all the stages of which were not waged with material weapons. It has been a war of equality against privilege, of change against the established order, of independence against reverence and authority. In spite of seemingly decisive victories, apparent success at the outset was rather superficial. The real heart of the nation had not been completely won over and convinced. And so the tide of battle flowed backwards and forwards; the banners of democracy now advanced, and now receded. But it will be observed that in each forward movement some ground was gained which was not lost in the ensuing retreat. Thus the Republic hastily proclaimed in 1870, though on several occasions in imminent peril, has been preserved almost as by a miracle, and appears stronger after every crisis. The victory is not yet complete, but the present outlook offers slight encouragement for the reactionary forces.

The National Assembly, under the patriotic direction of Thiers, regulated the finances, paid the huge war imdemnity with unexpected quickness, and showed the world that France was not moribund. But a majority of this assembly was made up of monarchists, and only their dissensions afforded the Republic a chance to live. Thiers was succeeded as president by Marshal MacMahon, in 1873, a pronounced royalist, who conceived it his duty to prepare the way for the advent of a king. A fusion of the Legitimists and Orleanists supported the claim of the Count of Chambord, as Henry V, to the throne of France. The romantic, but petty devotion of the Count to a symbol brought about the failure of the royalist design and the salvation of the Republic. The Orleanists stipulated that the tricolor flag be retained as recognition of the more fundamental changes wrought by the revolution; but Henry V could not be unfaithful to the white flag of Henry IV. The Bourbons—at least the elder house—could not forget.

The Assembly grudgingly drew up a constitution in 1875, republican in substance, although scarcely avowing itself formally as such. The Senate of three hundred members was established, who are elected for nine years by electoral colleges, of which there is one in each department, composed largely of delegates from the communal councils. The Chamber of Deputies was to be elected by direct, universal suffrage for a period of four years. The two houses together constitute the National Assembly which has power to amend the constitution, and elects the president for a term of seven years. The ministers were declared to be jointly and severally responsible to the chambers for the general policy of the government. Republicans were a majority in both chambers from 1878, and MacMahon resigned the following year.

The second great crisis through which the Republic had to pass came in 1888, when General Boulanger, some time Minister of War, was cultivating the engaging arts of Prince Louis Napoleon, later Emperor Napoleon III, with a view to emulating his example. The Republic had not yet won general confidence. The clerical and military circles were largely hostile to it. The belief was quite generally entertained that the unstable republican régime was a military weakness. France was still isolated in the face of the

Peace terms, 1871: Bismarck dictating conditions to Thiers. *After the painting by Carl Wagner.*

Surrender of Sedan. *After the painting by Anton von Werner.*

greatly superior power of Germany. This fact was the cause, undoubtedly, of nervousness in French policy. Boulanger made known his program of rather vague proposals calculated to attract discontented persons of various parties, direct election of the president being the most plausible feature since it pretended to strengthen the popular influence. A summons to appear before the High Court to answer to charges of conspiracy and treason awaited Boulanger after he had been elected deputy, and he took refuge in Belgium, and died there. Soon after this, the intimacy and alliance with Russia added assurance to the Republic, and as it felt itself firmer, it became more radical.

But there were still dangerous elements unwilling to acquiesce in the Republic, although it had proved its vitality by living longer than any previous government since 1789. The opposition to the Republic was without a rational foundation. Under whatever name it raised its head, it aimed to set up a principle of government which had repeatedly failed ignominiously. It is not to be supposed that any considerable number of intelligent persons continued to support this opposition. It was an alliance of simplicity, sentimentalism, and dishonesty; of the timid, of those whose emotions were deceived by obsolete shibboleths, and of leaders who disguised the true nature of their designs by lofty expressions of piety and patriotism. The true purpose of the opposition was the exploitation of the many by the few.

The recent history of France exhibits with exceptional clearness the dangerous currents in military and ecclesiastical circles, and the policy of the government regarding them. In the celebrated Dreyfus Case, although nominally one man alone was involved, the forces of reason and tradition were really arrayed on opposite sides. A contest was carried on for more than ten years with the utmost zeal of enlightened individuals to raise justice above artifice, prejudice, partisanship, and the irrational spirit of caste exclusiveness, which tends to develop in military establishments and to make itself invulnerable by the assumption of a sacred, inviolable character.

Captain Alfred Dreyfus, an Alsatian Jew, was arrested in October, 1894, on charges, submitted by one Major Esterhazy, of selling military secrets to a foreign power, and condemned by a court-martial the following January to public degradation and deportation for life to an unhealthy island off the coast of French Guiana. Later, Colonel Picquart became convinced that the document which had served as proof of the guilt of Dreyfus was Esterhazy's own forgery. The military authorities rid themselves of Picquart's inconvenient presence by sending him for service in Tunis and Algeria. Later he was imprisoned on charges preferred by Esterhazy. Then the novelist Zola, who had become convinced of Dreyfus' innocence, wrote for publication, in L'Aurore, a scathing letter addressed to the president. He was condemned to imprisonment in his absence, having crossed over to England as precaution. It is needless to say that each successive incident increased the publicity of the case, and nourished popular excitement. The foreign press alluded to this travesty of justice, as affording convincing proof of the moral decadence of France.

General Cavaignac, Minister of War, now produced three new documents, and persons asked themselves, why, if Dreyfus had been rightly condemned, it was necessary to have recourse to this cumulative evidence of his guilt. Picquart declared that two of the new documents were irrelevant, the third a forgery, and very soon a Colonel Henry confessed to this forgery and shot himself. Cavaignac asserted that this incident did not affect the earlier trial; but it is obvious that its profound effect upon the general opinion concerning that trial was inevitable. The Court of Cassation, on June 3, 1899, ordered a new trial of Captain Dreyfus. This trial was conducted by a court-martial at Rennes with popular excitement at a fever heat. The trial was a mockery of justice. It was evident from the beginning that the officers who composed the court had agreed that a condemnation was necessary to maintain professional prestige. The defendant was found guilty with extenuating circumstances, and the President of the Republic pardoned him. The court condemned themselves by this condemnation; for how can there be extenuating circumstances for treason? Finally, on July 12, 1906, the Court of Cassation, after reviewing the case, annulled the verdict of Rennes and declared all the incriminating documents to be forgeries. Captain Dreyfus was promoted to major, and received the decoration of the Legion of Honor. Picquart's service to justice was appropriately rewarded when he became Minister of War. France was finally vindicated. The public had been torn with discord; but the successful issue of this long struggle added strength to the Republic. The accusation of Captain Dreyfus was part of an anti-Semitic movement in which some of the clerical party, and other reactionary groups were engaged. But the trial was also the battle-field for other issues, particularly for a contest against military arrogance. It was necessary for the progress of the spirit of democracy in France that any arbitrary or partisan tendency in military circles should be restrained.

The beginning of a new era in the history of republican France was the formation of the Waldeck-Rousseau Ministry in 1898. Up to this time the executive had suffered from the paralyzing influence of political chaos in the Chamber. There had been no year since the establishment of responsible government which had not seen a change in ministry. The Waldeck-Rousseau Ministry reposed upon a *bloc*, or combination, of all the more progressive groups in the Chamber. This radical alliance, to which Socialists have at times been admitted, has endured to the present. Its existence, by the very fact that the other groups have united their forces in opposition to it, has tended to bring about a two-party division of political strength, which is essential to the parliamentary form of government. The *bloc*, under the leadership of a number of vigorous statesmen, has left a remarkable record of reforming legislation, the most striking feature of which has been the separation of Church and State. France, too, seemed more absorbed in the problems of internal improvement than in the more illusory aims of foreign policy, in 1914.

We are mainly interested in Russia's foreign policy, because by it she became involved in an inevitable conflict which is one of the positive causes of this great war. But some acquaintance with her recent internal development and problems is indispensable for an intelligent appreciation of her foreign relations and the motives of her diplomacy. We may profitably consider in this connection the political development which culminated in the granting of the constitution, the economic situation of the peasants and agriculture, and the progress of industry and commerce.

It will be recalled that the reign of Tsar Alexander II (1855–1881) was ushered in with a series of amazing reforms affecting directly and profoundly a larger number of souls than any similar innovations recorded in history. Forty-six million serfs were liberated from bondage, the judicial system was transformed according to western European standards, and elective provincial councils were established by means of which popular influence was introduced into the local administration. But during the latter part of the reign the imperial reforming activity was dis-

Joseph Chamberlain. His political career began about 1868 and from that date he held many offices, being prominent in parliament and the cabinet. He died July 2, 1914.

continued, and the radical element in society awaited in vain what they expected as the crowning act, the conferring of a constitution. Impatience generated agitation. Secret societies were formed, especially among the university students. At that time the elements in Russian society which laid exclusive claim to the distinction of intelligence, the "intellectuals," suffered largely from a sort of mental fever caused by the imperfect assimilation of the elements of western learning and culture upon which their minds had been gorged. Those who formed this class looked upon sociology as the most distinguished and beneficial of the sciences, and in their inexperienced eagerness for social improvement they brought forward many impractical and foolish schemes. The programs of the extremists were subversive of society. Nihilism was a revolt not against this or that institution, but against the whole established order. Agitation had recourse to terrorism, and on March 13, 1881, the Tsar Liberator was assassinated in St. Petersburg.

The reign of Alexander III (1881–1894), the son and successor of the murdered Tsar, was a period of reaction, and of the repression of all tendencies towards liberalism. The Tsar stood for the resolute maintenance of the undiminished principle of autocracy. Besides, the more substantial elements in society had been disgusted with the extravagant vagaries of the extremists in social reform. Under Alexander III essentially national Russian ideals were cultivated in distinction to those which came from western Europe. Pan-Slavism was in high favor, which emphasizes the unity of the Slav race and idealizes their distinctive spiritual gifts and possessions. Pobyedonostseff was the chief adviser of the Tsar, a man of positive, narrow convictions, intensely hostile to the liberal ideas of western Europe.

Nicholas II, son of Alexander III, announced at the time

of his accession his strict intention of following in his father's course. Nihilism had disappeared, and the more violent forms of agitation had largely subsided; but the desire for constitutional government gradually pervaded all classes. Pobyedonostseff's influence was powerful until 1904, but Sergius de Witte, who was appointed Minister of Finance and Commerce in 1892, believed that the experience of western Europe indicated the road to true national prosperity.

The Trans-Siberian Railway, begun in 1891, was completed in 1902, linking together with bonds of steel the extremities of the empire. But the construction of this line was part of a plan of expansion which led to the occupation of Manchuria, and the acquisition of Port Arthur. This brought on the war with Japan. The contest, which was generally unpopular from the first in Russia, inflamed the indignation of the people more and more by the spectacle of administrative incompetence and corruption and the succession of disasters with which it was attended. The war began in February, 1904, and the assassination, July 28, of harsh, inflexible von Plehve, Minister of the Interior, released, as it were, a mighty volume of discontent. As the war progressed, this current swelled. The fall of Port Arthur was followed by Red Sunday, January 22, 1905, when the Cossacks fired a volley into the crowd before the winter Palace in St. Petersburg. The Zemstvos, or provincial councils, sent delegates to a succession of congresses where the necessary reforms were freely discussed, and the demand for parliamentary institutions formulated. The termination of the Russo-Japanese War by the Treaty of Portsmouth, September 5, did not relieve the domestic situation, but it released the enlightened Count Witte from his duties as envoy, so that he returned to Russia and soon became Prime Minister in place of the hated Pobyedonostseff.

Russia was seething with discontent. Acts of violence occurred daily. A general strike, carried out in October as convincing protest against the hesitancy of the government, began with the railways and extended to the other occupations. For a time the activity of almost the whole country seemed to stand still. The powers of opposition yielded, and on October 30 the Tsar issued his famous manifesto, in which he promised the fundamental rights of civil liberty, and the participation of an assembly of representatives of the people in acts of legislation.

The elections for the first Duma, or parliament, took place in March and April, 1906, on a very democratic basis of suffrage. On May 5, de Witte resigned, and Stolypin took his place. The next day, the "fundamental laws" were issued by autocratic decree, limiting the scope of the popular concession by instituting the Council of the Empire, as a kind of upper chamber, to which the legislative bills of the Duma were to be submitted. One-half of its members are appointed by the Tsar, the other half are elected by the Zemstvos and other corporations. The first two Dumas were short-lived, sitting from May 10 until July 22, 1906, and from March 5 until June 16, 1907, respectively. They were both dissolved by decree of the Tsar. Their demands, such as for ministerial responsibility, and the immediate distribution of the crown lands to the peasants, seemed excessive. It was clear that no Duma chosen on so broad a suffrage basis could coöperate harmoniously with the government. Stolypin was putting down the lawlessness that survived the revolution with uncompromising severity. The government recovered its assurance, and after the dismissal of the second Duma, modified profoundly the voting system by decree in open violation of the constitution, substituting a very complicated method of indirect elections, with a sort of combination of the Prussian timocratic classification with the native Russian division of society as basis. The result was that the influence of the masses in the composition of the Duma was reduced to a minimum. The third Duma was naturally much more conservative, although not always as submissive as had been anticipated. Its period of existence attained the normal limits, from 1907 until 1912. The restriction of the basis of suffrage was unavoidable, no doubt, in a country where three-fourths of the population is illiterate. The prevailing ignorance of the masses is commonly represented —and rightly—as a serious obstacle to the progress of the nation, particularly in a political sense. It is only fair to note, however, that the educated classes, who alone exercise any appreciable influence in the government of the country, compensate somewhat by their zeal in learning for the ignorance of the great majority of their compatriots. For, while 12,967 books were published in the United Kingdom in 1913, and 12,230 in the United States the same year, the annual publications in Russia amounted to 29,057 in 1910, the most recent year for which returns are available, and increase at the rate of about 3,000 yearly. One may be safe in assuming, therefore, that the 43,000,000 literate Russians read more books than all the 150,000,000 literates of the Anglo-Saxon race throughout the world.

At Stolypin's initiative, valuable work in constructive legislation was accomplished; although, unfortunately, he encouraged, for support in carrying out his program, the development of a narrow, aggressive, nationalist party, hostile to the privileges of the subordinate nationalities, as Poles and Finns. In the Fourth Duma, elected in 1912, the conservative elements have a large majority.

Popular rights are still limited in Russia, and the existence of parliamentary institutions is somewhat precarious. The recollection that the embarrassment of the Russian government

Count Serge Julievich Witte, president of the first ministry under the new Russian constitutional régime.

Peter Arkadievich Stolypin, Russian Minister of Interior, 1904, and Prime Minister, 1906.

in the war with Japan had been seized upon immediately as an opportunity for revolution, the consideration that this movement had been repressed before reaching its intended goal and that therefore much bitter disappointment must lie hidden beneath the surface, and, finally, the observation that labor troubles were numerous, and that in some instances, at least, their motives had not been without some connection with political agitation, might altogether have supported the conviction as late as 1914 that a formidable hostile action against the empire would set in action a second time the forces of disintegration, which would facilitate enormously the progress of an invader.

Agriculture is the fundamental occupation in Russia, and the empire is in large measure a peasant empire. The prosperity of the empire advances or recedes in accordance with the progress or decline of the agricultural interests. Even now that an important industrial class has been created, of every 1000 inhabitants of the Russian Empire, 771 are peasants, while 107 belong to the middle class, 23 are Cossacks, 15 are nobles, 5 are ecclesiastics, and the others belong to other minor classes. Whoever would know Russia must make the acquaintance of the peasants. The character of the Russian peasant is variously judged. Some writers have shown a tendency to idealize his primitive, simple environment, and his direct, but often quaint, manner of thought and expression. A native shrewdness and good-sense is claimed for him. But the opinion is also common that he is submerged in ignorance, sloth, and drunkenness. The proportion of illiteracy is highest, of course, among the peasants. Climate and natural environment have probably contributed somewhat to make the Slavs of the great plain less alert and strenuous than the peoples of western Europe. Statistics go to show that the evil of intemperance is not quite so extreme as has been supposed.

While in late years the Russian has consumed 2.47 liters of spirits annually, the German consumption per capita has been 4.1 liters.

Upon obtaining their personal liberty the former serfs continued generally in possession of the lands which they had been cultivating for themselves before the act of emancipation. The value of the customary services which they had rendered annually to their former proprietors was capitalized at 6%, and this capital sum was reckoned as a redemption debt which the peasants must eventually pay. To facilitate the operation, the government turned over at once four-fifths of the sum to the proprietors, establishing as method of liquidation that the peasants should absolve this debt and interest together by forty-nine annual payments, each equivalent to 6% of the sum thus advanced. The peasants were grouped in small, self-governing communities, which Russians designate by the term *mir.* The land cultivated by the members of the individual *mir* was owned collectively by the community. It was commonly redistributed from time to time among the peasant households; but as a fundamental division of all the land into three large portions was first made to serve as a basis for the annual rotation of crops, each family received its particular allotment in detached strips, part in each portion of the communal domain.

In 1905, of the land of European Russia, 40% belonged to the state, 35% to the peasant communities, and 25% to private proprietors. In the peasant communities there was an average apportionment of 27½ acres for each household. This, with the generally primitive, unscientific methods of agriculture, was insufficient, particularly in view of the prevailing generous size of peasant families. The organization of the *mir* was too rigid to respond to the requirements of the rapidly increasing population. Its territorial limits had remained stationary, and therefore, with every redistribution, the amount of the individual allotment had diminished. The situation reached a crisis in 1905. If the government had been assured of the loyalty of the peasants, it might have rejected the demands of the urban population. But the peasants were in a violent ferment, and even the constitution brought them no direct advantage.

The government finally undertook to relieve the situation in three ways, by dissolving the *mir* and substituting private ownership, by aiding the peasants to enlarge their holdings, and by encouraging emigration to Siberia. From 1907 to 1912, the total emigration to Siberia amounted to 2,400,000.

The buoyancy with which Russian finance responds to favorable agricultural conditions has been exemplified after the restoration of political order by the effect of two excellent crops in 1909 and 1910, and a medium harvest in 1911. The receipts of the treasury increased at once and industry and commerce were greatly stimulated. The increase of exports from 1907 to 1911 was 51%, and of imports, 37%.

The plain extending from the Dnieper to the Volga, and even beyond, in southern Russia, is called the Black Earth Belt from the character of its incomparably fertile soil. This soil, varying in depth from one to twelve feet, is capable of yielding excellent crops many years in succession without artificial fertilization. Likewise in the south, in Little Russia, the cultivation of the sugar-beet is carried on very extensively.

Russia is capable of becoming the granary of the world. Her crops in 1912 amounted to 727,011,000 bushels of wheat, 1,067,584,000 bushels of oats, 1,043,982,000 bushels of rye, 464,124,000 bushels of barley, 1,176,055,000 bushels of potatoes, and 1,808,800 tons of beet-sugar. Moreover, intensive agriculture has scarcely gained a footing, machinery is only partially employed, and there are vast areas of virgin soil still awaiting the plow.

The rapid development of industry in Russia began with the appointment of Count Witte as Minister of Finance and Commerce in 1892.

His policy was to place the currency on a gold basis and make the empire industrially self-sustaining. To promote the industries of Russia, he persuaded the government to establish very high duties on imported articles. He also did everything in his power to attract foreign capital to Russia. Not only was foreign support necessary for building up the infant industry of the country, but the influx of gold would contribute to the success of his currency reform. Witte succeeded in establishing the gold standard in Russia, which is his greatest single contribution to the welfare of his country.

Unconsciously the Russian government coöperated by its high tariffs in developing the forces of opposition. For the rise of industry resulted in the formation of a numerous laboring class in the large cities possessing greater intelligence and solidarity of sentiment than the peasants. This factor was of decisive importance in the revolutionary movement of 1905.

The first cotton factory was built in Russia in 1840, and now over 7,000,000 spindles are in operation spinning about 2,000,000 bales of cotton annually. The Russians are endeavoring to make this industry independent of foreign sources of supply of the raw material by developing cotton-raising in Turkestan, where it is already a flourishing occupation.

The center of commercial gravity in Russia is moving southwards in the direction of the most productive natural sources of wealth. An immense bed of coal, apparently inexhaustible, has been discovered in the basin of the River Donetz, near its junction with the Don, northeast of the

Sea of Azoff. As iron and coal are found here in proximity, nature seems to have equipped this region especially for becoming a bee-hive of industry. There are also valuable coal-fields in Poland, which has developed rapidly as a manufacturing region.

The traffic in alcohol and spirits was made a government monopoly in Russia in 1894. The government purchased the product from private distillers and undertook the distribution and sale. The receipts of this monopoly in 1913 were about $435,240,000, the expenses $116,480,000, leaving a net income of $318,760,000, a handsome profit.

The annual production of Russian manufacturing enterprises was valued at approximately $2,670,000,000 in 1913. The production of coal had only attained about 30,000,000 tons before the war. The annual production of 68,019,208 barrels of crude petroleum was nearly equivalent to that of the United States in 1902. In pig-iron, 4,131,248 tons, Russia's record was about that of the United States in 1881.

European Russia is so nearly flat that its great rivers are very sluggish, and therefore navigable almost to their head-waters. The entire descent of the Volga, for instance, throughout a course of about 2,400 miles, is only slightly more than 800 feet; and a project has been elaborated for a ship-canal from the Baltic to the Black Sea with only two locks. The internal water-ways of Russia in Europe are navigable to the aggregate extent of 20,670 miles by steamer, those in Asia, 21,421 miles. Two thousand river steamers ply on the Volga alone. These and many of the locomotives consume as fuel crude petroleum from the oil-wells in the Caucasus.

In 1913 there were 46,839 miles of railway lines in Russia. She holds second place in this respect among the great nations. Of the railway mileage, 60% is owned by the state, and the government lines produce a net annual revenue of about $158,000,000.

Russian exports in 1913, valued at $823,264,000, were slightly ahead of those of the United States in 1895; the imports, valued at $684,684,000, were slightly above the American record of 1894.

The exports of cereals attained the value of $368,082,520 in 1911. The Russians produce rye principally for their own consumption, but wheat forms the most extensive commodity for export; and without her exportation of grain, Russia would be unable to meet the annual interest on some $5,000,000,000 of foreign capital invested in the empire.

An approximate summary of the principal headings and total receipts and expenditures of the budget estimates for 1914 will illustrate the condition of the imperial finances, before extraordinary war expenses modified the customary equilibrium.

Summary of Russian Imperial Budget reckoned in United States currency in 1914.

Receipts		Expenditures	
Direct taxes.......	137,280,000	Finances	256,880,000
Indirect taxes	368,680,000	Public instruction..	84,240,000
Monopolies	555,880,000	Routes of communication	384,280,000
State property.....	579,280,000	War	311,480,000
Other sources......	189,800,000	Marine	130,520,000
		Debt.............	209,560,000
		Other expenses	340,080,000
	$1,830,920,000		$1,717,040,000
Extraordinary sources of income	7,280,000	Extraordinary expenses.......	133,120,000
Deficit	11,960,000		
Total	$1,850,160,000		$1,850,160,000

The total is surely a colossal sum, in keeping with the bigness of Russia. It may be observed that the expenses are counterbalanced mainly by the receipts from monopolies (net income from alcohol and spirits alone $318,760,000) and state property (railways, crown lands, etc.), so that the burden of direct taxation for imperial purposes is very light. The ordinary expenses for the military and naval extablishment is equivalent to about $2.52 per head of the population of the empire, as compared with $5.08 in Germany, $7.72 in Great Britain, and $6.50 in France. The indebtedness of the Russian government amounted in1911 to a little more than $4,500,000,000. The Bank of Russia, which alone issues paper currency, is virtually a government institution. Its policy has been commendably conservative; for of late years the currency of the bank in circulation has been actually less than the value of the bullion stored in its vaults. Therefore, it was possible to finance the war in its early stages chiefly by expanding the paper currency, and still keep well within the danger line. Thus, on September 14, 1914, the currency in circulation was equivalent to $1,327,560,000, whereas the bullion in possession of the bank was $958,880,000. On the basis of this reserve, the largest agglomeration of the precious metals in any repository in the Old World, the bank could easily have expanded its circulating medium another $1,000,000,000 without attaining the stage of recklessness, according to western European practice.

There were these numerous indications that the Russian Empire was entering upon an era of brilliant, sound prosperity. Nevertheless, the change had been so recent, and the influence of favorable tendencies appeared to be still so superficial, that many German observers viewed the situation with skepticism. General Bernhardi, in his work *Germany and the next War*, declared that the whole body of the Russian nation was so tainted with revolutionary and moral infection, and the peasantry was plunged in such economic disorder, that it was difficult to see from what source a vivifying force could spring up capable of restoring a healthy condition.

We have noticed how the action of the intense, concentrated energy of Prussia in achieving the unification of the Fatherland inevitably transformed the spirit of the rest of Germany, assimilating it largely to the character of the more robust portion. Heinrich von Treitschke, the famous professor of history in the University of Berlin, declared that the impartial observer must be convinced that "since the Great Elector the political history of Germany is entirely contained in Prussia," and that every "clod of land which was lost through the fault of the old Empire, and was won back again, was acquired by means of Prussia. In that state," he asserted, "lay thenceforth the political energies of the German nation just as certainly as she did not contain in her for a long time the ideal energies, in fact almost thrust them from her."

The doctrines of the great Prussian political teachers are a conspicuous illustration of the purely relative value of all precepts of political philosophy, inasmuch as the institutions, which they emphasized as universal standards were precisely those of which Germany had stood in need. They were the institutions by virtue of which Prussia had grown strong, and which the other states accepted in adopting the Prussian organization as a model. Chief among these institutions was universal obligation to military service. The national army had been a sort of school for implanting the qualities most useful for German citizens. The Germans had been disunited, irresolute, visionary. They were trained to coöperate, to be steadfast and practical. The directing and intellectual forces in Prussia seem to have collaborated with conscious unity of purpose in developing the cohesion and vigor of the nation.

There is a sense of national exhilaration and expansion of

David Lloyd George, Chancellor of the Exchequer, Great Britain.

Herbert Henry Asquith, Prime Minister and First Lord of the Treasury, Great Britain.

spirit in the harmonious action of a comprehensive organization such as the German military establishment. It is the nation in arms. Naturally the Germans regard as a defect in any other nation the lack of this national organ of common effort and efficiency.

In 1888, when at Bismarck's proposal the potential strength of the forces for war was increased by adding to the active reserve the Landwehr and Landsturm, the Chancellor affirmed that by this measure the German army became an even stronger guarantee of peace. "As matters stand," he said, "no war would be possible unless it were approved by all those who are liable to service, in other words, by the entire nation; and such a war would be conducted with the enthusiasm of 1870, when we were the object of an insolent attack. Under such circumstances all Germany would bristle with guns, and woe to the rash foe who should presume to defy the fury of the Germans (*furor Teutonicus*). I frankly warn foreign powers to discontinue their repeated menaces; we Germans fear God, but nothing else in the world." Bismarck's "blood and iron" theory of Prussian and German evolution was reëchoed by William II in addressing a gathering of officers at the palace in 1891. "The soldier and the army," said he, "not parliamentary majorities and decisions, have welded together the German Empire."

A somewhat detailed examination of the German Constitution may assist us in weighing for ourselves the measure of truth in the widely divergent views which have been expressed concerning its character. The Imperial Constitution is doubtless the creation essentially of Prussian statesmanship. It is a continuation of the instrument of association of the North German Federation with the modifications introduced in consequence of the union with the South German States. The Empire includes twenty-six states, and the sovereignty resides in the Bundesrat, or Federal Council, an assembly of the representatives of the governments of the different states. There are sixty-one members, or more strictly votes, in the Bundesrat, and their distribution among the states varies considerably with a tendency to correspond to the importance and size of the individual divisions, but without attaining a true proportionate adjustment. Thus Prussia has seventeen votes in the Bundesrat, while seventeen of the states have one vote each. The presidency of the federal government belongs to the King of Prussia, and he bears the title Deutscher Kaiser, or German Emperor, by virtue of the exercise of this function. The Kaiser declares war and makes treaties and other agreements in the name of the federation; but only in case the territory of the Empire or its coasts are attacked, is he empowered to declare war without the consent of the Bundesrat. He appoints the Reichskanzler, or Imperial Chancellor, his chief minister, who is one of the Prussian members of the Bundesrat. The Chancellor presides in the meetings of the Bundesrat, appoints the federal ministers and other officers in the name of the Emperor, and supervises them in the performance of their duties.

The Bundesrat adopts such general administrative measures as are necessary for putting the laws into execution. The prominent position of the Bundesrat does not depend solely upon its executive prerogatives as repository of the sovereignty. It exercises a controlling influence in the legislative function, which it shares with the popular organ of the government, the Reichstag. The Reichstag consists of 397 members elected by the universal suffrage of all German citizens who have completed their twenty-fifth year of age. The original design was that the voting districts, returning individually a single member, should contain each 100,000 inhabitants. But no re-apportionment of the districts to keep step with the increase and movement of the population has ever been made; and, in consequence of this, the great industrial centers, which have grown very rapidly during the past generation, are now inadequately represented. Proposals for new legislation are usually first made in the Bundesrat. They are then submitted to the Reichstag by resolution of the more august body. Members of the Reichstag may also present proposals for legislation; but such bills after being voted by the popular chamber, must receive the final assent of the Bundesrat. The Kaiser does not have the power of veto in legislative matters, because the sovereignty is vested in the Bundesrat, which derives its authority from the sovereign states of which the federal union was composed. The members of the Reichstag represent the German people as a whole, while the members of the Bundesrat represent the governments of the individual states. The Reichstag is dissolved every five years. It may be dissolved oftener by decision of the Bundesrat with the Kaiser's consent. The life of the Bundesrat is, however, by its very nature, continuous.

The Bundesrat is sometimes looked upon as the upper chamber of the legislature, like the Senate, or House of Lords. The analogy is far from exact, a fact which is illustrated by the circumstance that the Reichstag alone is frequently called the Parliament. The Bundesrat is not, in the fullest sense, a deliberative body. Its members are merely agents of the state governments which send them. The delegation from each state not only casts its votes as a unit, "votes solid" in other words, but it is under obligation to vote according to the instructions which it receives from the government which it represents.

It will be observed that the cabinet system of government has not been established in Germany. By "cabinet system" is understood the adjustment of powers which prevails in

England, and the European states which have imitated more closely English parliamentary institutions, by virtue of which the ruling power is exercised by a ministry chosen from the prevailing party in the popular legislative chamber, or in any case so sensitive to the opinion of that house, that it resigns as soon as it ceases to enjoy the support of the majority of its members. The German government does not admit this responsibility of the ministers to the people's representatives. The heads of the great departments of government are responsible to the Chancellor, he in turn to the Kaiser. Thus the Kaiser exercises greater personal authority than most of the heads of constitutional states.

Amendments to the imperial constitution would be enacted through the ordinary channels of legislation. Amendments calling for an enlargement of popular influence in the government would presumably be rejected by the Bundesrat, where Prussia could enforce her disapproval; since fourteen votes alone can defeat a constitutional amendment.

As the influence of Prussia is so decisive, a few words about her own political organization are necessary, to show why the policy of her government is quite uniformly conservative. The Prussian ministers are responsible to the king who appoints them without reference to popular opinion. The legislature, called Landtag, consists of two chambers, the Herrenhaus, or House of Lords, and the Abgeordnetenhaus, or House of Deputies. The privilege of membership in the upper house is conferred by the king. The members are either hereditary peers, or life peers. The lower house is made up of four hundred and forty-three members. The method of electing these deputies is the cause of serious popular discontent, and the object of bitter ridicule. It is probably the greatest obstacle to the extension of the popular influence in the government, not only of Prussia, but of the empire. This peculiar character of

The Julius Tower in the citadel at Spandau, the Prussian fortress. In the tower is popularly supposed to be kept the German parliamentary institutions in Prussia is due to the fact that the constitution was granted in 1850, during a period of reaction which followed the liberal movement of 1848–9. Those who drew up the constitution proposed to satisfy a popular demand by establishing the principle of universal suffrage, but at the same time to make it practically ineffective by the manner of electing the representatives. They devised what is called the "three-class system" of voting. The voters in each constituency, all men who have completed their twenty-fifth year, are divided into three classes on the basis of the taxes which they pay to the state, in such a way that the aggregate sum of the taxes paid by the voters in each of the classes is equal, or in other words, equivalent to one-third of all the taxes contributed by the district. The voters are apportioned to the three classes in such a way that the largest tax-payers are assigned to the first class, the next group of tax-payers in the descending scale to the second, and finally the poorest citizens to the third. Each class in an electoral district elects, without regard to the number of voters which it contains, the same number of secondary electors, who are called Wahlmänner. The number of Wahlmänner for each district is determined by the ratio of one for two hundred and fifty of the inhabitants. These secondary electors choose the representative to be returned to the lower house of the parliament by the district. The system is illogical and unjust. It is obviously not a uniform timocratic basis for the division of the voters into classes throughout the constituencies of the kingdom. For there is, of necessity, a bewildering variation in the property range of the classes according to the amount of wealth and the manner of its distribution in each of the districts. There are districts where a single tax-payer constitutes the first class; and a curious instance is related of a man in Berlin who constitutes alone the first class in one of the poorer, eastern electoral districts of the city, but who, if he moved to a fashionable quarter near the Tiergarten, would be transferred in consequence to the third voting class. The Imperial Chancellor himself votes as a Prussian subject in the third class in the district where he lives. The injustice of the system is most conspicuous, of course, in the practical exclusion of the poorer, and most numerous, elements of the population from representation. As the secondary electors representing the two upper classes very often vote for the same candidate, it follows that the third class, although it embraces an overwhelming majority of the population, is generally excluded from all influence in the elections. The actual figures in recent elections show that the first class of voters, electing one-third of the Wahlmänner, or secondary electors, contains in the aggregate about 200,000 voters, the second class about 900,000, and the third class more than 6,000,000. The representative character of such a system is clearly illusory. The suffrage is practically limited to less than one-sixth of the citizens. The popular influence on the government of Prussia is confined to this small fraction, who represent the interests of property. The government of Prussia is the dominating factor in the Bundesrat by its seventeen votes, and the initiative in imperial legislation is practically confined to the Bundesrat, if not to the Imperial Chancellor and the Prussian delegation. The government of Prussia is not responsible to the Landtag, or Prussian Parliament, and therefore probably the small fraction of the Prussian people, whose influence is really effective in the Prussian elections, exercises no appreciable control through parliament over the federal, or imperial, policy of the Prussian delegation in the Bundesrat. It follows, therefore, that any popular control at all over the German imperial government is small. It is exerted chiefly by negative means. The really popular parties in the Reichstag are essentially parties of opposition. The question is frequently discussed, whether there is a movement in Germany in the direction of responsible, or parliamentary, government. Incidents will be mentioned later of the greatest interest in connection with this question. But the kernel of the whole matter can be briefly formulated. The independent position of the government is rendered possible, and even necessary, by the comparatively chaotic condition of political parties. Responsible government could be readily won simply by setting up what is at the same time the indispensable basis for its operation, a two-party political system. Bismarck explained this truth: "The parliamentary system works easily, and, so to speak, with elegance when there are but two parties, as in England, where there are only the Whigs and Tories." "If there were similarly with us a party composed of a majority, it would always be a pleasure for the minister in power to attach himself to it, if not openly, at least by a secret alliance, and thus to work in harmony with it. But we are still far from this ideal. We have here something like eight factions; the German holds strictly to the spirit of his party, and keeps aloof from others."

The fragmentary condition of parties contrasts lamentably with the marvellous capacity for coöperation displayed by the Germans in other fields. The internal organization of one of the parties, however, is a wonderful example of German system, thoroughness, and discipline, that of the Socialists. The government aims to keep itself in the clear atmosphere above the hazy confusion of party conflicts. In actual practice, however, it often descends to court the favor of individual groups or parties by timely concessions for securing a temporary majority. The Center, representing the Catholics, is possibly the greatest obstacle to a rational alignment of parties as in Great Britain, France, or Italy. It is an unsound feature of German political life that a compact party of about a hundred members holds the balance of power in the Reichstag whose fundamental motive for association is not political in its nature. Politically the Center is an opportunist party. Without this un-political party the other groups could probably all be arranged along the same line, so to speak, according to the relative nature of their views, from ultra-conservative, or right, through various gradations to ultra-radical, or left. If this could be done, a cleavage would be conceivable at some point, and the groups on each side of this intersection might be drawn together for the common support of programs composed of practical measures consistent generally with their particular views and without compromising their individual theories.

This summary of some of the provisions of the federal and Prussian constitutions and of the chief features of political activity will help to make clear the fact that while there is naturally a tendency to ascribe the policy of the German Empire exclusively to Prussia, the view has also been expressed that Prussia's supposedly warlike attitude is due to a single class, the Junkers, or aristocracy of great land-lords.

Large estates prevail in extensive parts of the north and east of Prussia. In fact, big estates include nearly one-fourth of all the agricultural land in Germany. It is said that in some of the more backward sections of the northeast the large proprietors often keep their laborers in a state of political tutelage. Tradition and environment have doubtless contributed to make the character of the Prussian land-owning aristocracy somewhat inflexible. The members of this class are very conservative and intensely loyal to the throne. Their attitude is often narrow, and not free from the suspicion of selfishness. They are given marked preference in the military and civil service, and enjoy almost a monopoly of the positions at court and in the diplomatic field. When we consider the way in which Prussian deputies are elected, and, in addition, the fact that the rural districts are manifestly over-represented in both chambers, we are not surprised that the Prussian Parliament is a stronghold of conservative and agrarian interests.

Current investigations of the responsibility for the war not only frequently charge Prussia with the blame, but discover within this kingdom the class or classes which are especially contaminated with the guilt. Now we are told that the Junkers precipitated the conflict so as to perpetuate their unjust oligarchical tyranny by diverting public attention from domestic abuses to the glory of military aggrandizement. Again we learn that the industrial and commercial class was the real instigator by its desire for colonies and additional foreign markets. In fact the intentions of the two are so frequently, and, as it were, so interchangeably invoked in this connection, that to bar an over-hasty inference that they were associated in a common warlike design, it is well to observe that the Junkers and industrials are naturally discordant elements. Their differences were revealed and accentuated by Bismarck's introduction of a protective tariff in 1879. This measure, it is true, aimed to promote the interests of both classes by encouraging German manufactures and by aiding to maintain agricultural self-sufficiency. But in placing duties on agricultural products the political intention was likewise prominent of consolidating the economic foundation of a class which is considered to be the chief prop to the Prussian throne. It can be shown that the duties as applied are almost exclusively favorable to the large proprietors.

The advance in food prices in consequence of the higher duties was a burden for the working classes, and affected indirectly the manufacturers by the resulting increase in wages and the cost of production.

Either the tariff or the enterprise and scientific methods

of the agriculturalists have accomplished wonders in Germany. Even since 1891 the most fundamental products of the soil show a remarkable increase, more than keeping pace with the population. (See Comparative Table of Statistics, page 79, Nos. 13, 14, 15, 16). The achievement of German agriculture, in view of the rather medium character of the soil, has been quite as remarkable, although less loudly proclaimed, than that of German industry.

The agrarian interests are the back-bone of the Conservative Party, as industrial and commercial interests are represented chiefly in the National Liberal Party. The Social Democrats alone form an adequate democratic opposition. Unlike the Radicals in England and France, to whom they in a measure correspond, they cannot hope to be represented in the government, unless an almost revolutionary change should occur. The Socialists suffer political, and in large measure social, ostracism. German Socialism aims to remedy the evils, the economic tyranny for the masses, resulting from the tremendous power given to private capital by the industrial revolution, that is, the introduction of the factory system of production. The Socialists' program as drawn up at Erfurt, in 1882, contains the following passage indicating their view:

"Only by transforming capitalized private possession of the means of production—such as land, mines, raw materials, tools, means of communication—into collective possession, and by making the production of goods a socialistic production carried on for and by society can it be possible to bring about a change, whereby the big industry and ever growing productive capacity of coöperative labor can become the possession of the classes hitherto exploited, and be transformed from a source of misery and oppression into a source of utmost prosperity and all-around harmonious improvement." Some of the particular reforms demanded by the

Socialists are universal, direct suffrage including women; popular initiative in legislation; a militia instead of the standing army; the absolute separation of church and state; graduated estate as well as income taxes, and the abolition of all indirect taxes.

Bismarck caused to be adopted very severe repressive measures against the socialistic propaganda; but proposed to give the people all the state socialism they wanted. Accordingly, he introduced the famous system of national compulsory insurance for all employes receiving less than 2500 marks yearly salary, to provide assistance for three distinct classes of disabled persons, (1) those temporarily ill, (2) those injured by accident, and (3) those invalidated through prolonged incapacity or old age. There are about 13,500,000 persons insured under this third heading, practically the whole laboring population and the government itself contributes to the amount of fifty marks annually for each pension. In this very important form of social legislation Germany has been the pioneer, and the United Kingdom has recently followed her example.

The Germans of to-day are characterized by patience, adaptability, a remarkable capacity for organization in both the active and passive sense, and thoroughness in detail. These characteristics are invaluable in an age when the great achievements are accomplished by applied science, a division of labor, and an intricate coöperation of many processes and activities. The character of the government and that of the people interact and are intimately related.

The qualities of thoroughness and discipline are exhibited in the work of legislation. A single note-worthy example will illustrate this, the enactment of the new civil code. Like other important legislative proposals it was first elaborated by a commission of experts. Frederic William Maitland praised this remarkable achievement of industry and thoroughness as follows:

"Among the great things that he (the German) has done is this; he has codified the greater part and the most important part of his law; he has set his legal house in order; he has swept away the rubbish into the dust-bin; he has striven to make his legal system rational, coherent, modern, worthy of his country and of our century." The commission worked on the project from 1874 until 1888, when it was published and freely criticized. It was referred back to a commission in 1892, and submitted in revised form to the Reichstag in 1896. Between January and July, 1896, the Reichstag passed this code of 2385 sections, and it was sanctioned by the Bundesrat, and went into force January 1, 1900. Maitland observed that there was less discussion in the Reichstag over this very lengthy, and tremendously important, act of legislation, than in the British Parliament over the passage of a recent (1902) Education Bill. This is a form of legislative efficiency which seems to be incompatible with party government. The periodic changes in government, and large absorption of energy in redeeming party pledges render impossible the sustained effort required for the creation of a supreme legislative monument of this kind in countries where the real sovereignty is not stable. The Germans have become by experience so convinced of the efficacy of coöperation, that they apply it to all important forms of activity, as may be perceived in the alliance of government and business, the intimate association of the banking and industrial interests, and the formation of syndicates. In fact, Germany is a vast society, all of whose activities seem to be nicely coördinated and adjusted.

There are eighteen railway systems in Germany, thirteen private, five state, and one imperial; but the public lines amount to 91% of the mileage. These are valued at $4,757,579,750 and produce a net income of $191,943,190.

The power to regulate the tariffs is vested in the Bundesrat, and exercised from time to time through a general conference of representatives of the different directorates. This is a powerful appliance for supporting German industry and commerce in international competition, for the central authority is empowered to grant special privileges, as rebates for large shippers, and exceptional rates for goods crossing the country or being conveyed to the sea-ports for exportation. In fact, the German state-owned railways employ the same devices which are loudly condemned when practiced by private lines in the United States. In Germany this method is accepted as part of a system carefully and impartially planned with a view to promote the national welfare.

A species of economic unification of the country has been accomplished through the intimate correlation of the banking and industrial systems. About seven large Berlin banks form the nucleus of this organization. Such institutions own shares sufficient to give them a paramount interest in many provincial banks, so that altogether they can control the greater part of the effective banking capital of the country. They participate, likewise, in most of the great industrial enterprises by owning considerable blocks of shares in the companies, and having representatives on the governing boards.

Most all of the great branches of industry in Germany have eliminated wasteful competition by constituting syndicates to control prices and regulate individual output. These organizations have been powerful engines for the expansion of German exports. At one time, when an incipient agitation against them led to an official investigation, a report was made that their destruction would curtail the ability of the country to compete abroad. Not only are the freight rates on the railways favorable to the syndicates, but the government has entered into relations with them in a manner which may lead to far-reaching consequences. Thus the potash syndicate, which was constituted in 1879, included in its membership the governments of Prussia, Anhalt, and the Reichsland (Alsace–Lorraine) by reason of the mines of which each of these states was the owner. This syndicate was dissolved in 1909, upon expiration of the contract which had been the basis of association, because one member believed that it would be more advantageous for him to operate independently. Then, at the instance of the Prussian government, the Reichstag voted a bill establishing a compulsory syndicate, a startling innovation. It was urged that the potash industry was intimately connected with the welfare of the country, was in short a sort of public utility, and must be guarded from demoralization. Yet one may readily predict that this will be the first of many statutory syndicates. The Prussian fiscus extracts about one-fourth of the coal mined in Silesia, and more than one-half the production of the Saarbrücken district. The Prussian government is not a member of the coal syndicate; but it will probably obtain in future a vote in this, as well as other similar important combinations. The Socialists do not view these developments with dissatisfaction. The formation of syndicates, the nationalization of the railways, the encroachment of the government upon the field of industrial development, all these are interpreted by the Socialists as necessary steps in the normal progress of society. Small concerns must inevitably be merged in great organizations, and they in turn, as a safeguard against industrial feudalism, must become public. We are tempted to adopt the opinion that in spite of the political animosity which separates the government and the Socialists their points of view differ mainly in the fact that the attitude of the latter, as being more radical, and less practical, is about a generation in advance of that of the former. By means of the inter-relation of great interests which has been described, it is said that about fifty captains of finance control absolutely the economic situation in Germany. Likewise, the general coöperation, and the partnership of government and business are carried to such a point that in spite of political dissension at home Germany presents an undivided front in the world-competition for commercial opportunity abroad.

The remarkable growth of German industry in recent times is too well known to require detailed description. The situation of the most important industrial districts has been determined largely by the position of the deposits of the minerals which are most fundamental for manufacturing. Coal and iron are found in proximity in Lorraine and Silesia, on opposite frontiers of the Empire. Westphalia is especially rich in coal; but the Silesian fields are said to exceed those of any other part of Europe in the probable duration of their supplies.

The table at page 79 has been arranged, partly with reference to the present chapter, to present in convenient form a comparative summary of the progress of Germany during a significant period of twenty years, partly as a useful collection of important economic data relative to four leading industrial nations, to be cited from time to time as the treatment requires it.

The reader will observe in consulting the table that in calculating the relative growth of exports, the percentage (No. 19) is based, not on the increase of the aggregate exportation of each nation, but on a comparison of the exportation per head of the population at the beginning and end of the period of twenty years. This seems more significant, especially in attempting to judge the influence of commercial competition in nourishing national animosity. Germany not only records a greater absolute increase in production than the United Kingdom, but in spite of the much more rapid growth of her population the rate of increase of the average share of each individual person in industry and commerce has advanced from 1891 to 1911 much more rapidly than in Great Britain (compare nos. 9, 11, and 19).

The progress of German industry is in large measure due to the practical application of the results of scientific inquiry. Striking examples of this are offered by the chemical and electrical branches. Artificial indigo was discovered in Germany in 1897, and was at once substituted for the imported vegetable indigo. In consequence, the manufacture of dye-stuffs from formerly useless by-products of gas and coke quickly rose to a production exceeding $30,000,000 annually. Reference has been made in another connection to the potash business. Germany has almost a monopoly of potash salts for fertilization, and the exportation of this commodity amounts to more than $30,000,000 annually. The development of the electrical industry during the last twenty-five years was made possible by the excellent technical schools. The annual exportation of electrical supplies amounts to about $40,000,000 in value.

The Franco-German War in 1870–1 left Germany without an imperial debt, because the expenses of the struggle had been covered by the indemnity exacted from France. By 1891 an imperial debt of $345,071,000 had been contracted, and its subsequent rapid increase to $1,224,808,000 during the next twenty years, a period of peace, and for unremunerative purposes, is looked upon in some quarters as an example of national prodigality. These figures alone do not afford an adequate basis for comparison with unitary states, such as France and the United Kingdom, where the central government performs the functions which lie within the field of activity of the different states in federal Germany.

Therefore, we ought to consider the aggregate indebtedness of the federal and state governments, which was $4,896,136,000 in 1911, as compared with the national debt of $3,567,498,000 in the United Kingdom, and $6,280,791,000 in France. The German debt is still less per capita than either of the others, and there is this striking element of difference in the situation, that the aggregate German debt is more than counterbalanced by the productive assets of the German states, primarily the railways. Germany, in this sense, had no *net* debt at the beginning of the war.

It is a well-known fact that after the termination of the war with France a quantity of gold, to the value of about $30,000,000 was deposited in a tower, known as the Julius Turm, in the fortress at Spandau about seven miles distant from the center of Berlin. This treasure was preserved as a military emergency fund, either to be actually used in case of mobilization, or to serve as additional security for issuing paper currency.

Some features of the economic situation are common, in greater or less degree, to the greater nations of western Europe, which are so fundamental in their relation to any phenomenon of actual life, that they should at least be mentioned. The resources of Great Britain, France, and Germany are not limited to their home territories or merchant flotillas. They are money-lending nations, which have invested vast quantities of capital throughout the world. The British investments outside the British Islands were estimated a few years ago at $13,000,000,000, those of France at $7,000,000,000, and Germany's at $5,000,000,000. British investments abroad are said to yield an annual return of 5.2%. British yearly savings for fresh investment are said to amount to about $1,500,000,000, French to $400,000,000, and German to more than $1,000,000,000. The annual increase in wealth in the United

States, by way of comparison, is variously estimated at from $2,000,000,000 to $3,000,000,000.

In becoming industrial nations they have ceased, to greater or less extent, to be self-sustaining in respect to food and raw materials for manufacture. The United Kingdom imports about 78% of its total supply of wheat and flour, Germany 35%, and France 3%. The United Kingdom depends upon outside sources for about 47% of its meat supply, and Germany for about 11%. Great Britain produces a surplus of coal, but finds it necessary to import more than a third of her supply of iron ore. France has an abundant iron ore supply but is dependent for about one-third of her coal consumption upon other nations. Germany produces a slight surplus of coal, and is nearly self-sustaining in iron ore.

Since the possession of large deposits of iron ore is a very important factor in industry, and a source of strength in warfare, it may not be inappropriate to conclude this chapter with a summary statement, in tabulated form, of Germany's relative position in respect to the possession of these supplies:

Estimate of Developed Supplies of Iron Ore in Tons.

Germany including Luxemburg	3,878,000,000
Great Britain	1,300,000,000
France	3,300,000,000
Whole of Europe	12,032,000,000
America	9,855,000,000
Australia, Asia, Africa	521,000,000
World	22,408,000,000
Germany's share of European Supply	32%
Germany's share of World's Supply	17%

A COMPARATIVE TABLE OF STATISTICS

For the Finances, Industrial and Agricultural Production, Commerce and Means of Communication of the leading Industrial Nations

Note:—Areas are given in square miles, money values in dollars, mineral production in long tons, and agricultural production in bushels.

		Germany		United Kingdom		France		United States	
		1891	1911	1891	1911	1891	1911	1891	1911
1	Area	209,793	209,793	121,633	121,633	207,054	207,054	3,574,658	3,574,658
2.	Population	49,428,470	65,186,000	37,879,285	45,370,530	38,343,192	39,601,509	62,622,250	93,402,151
3.	Area of colonies and dependencies	966,150	1,026,022	9,059,851	11,371,437	2,814,988	4,538,543	....	115,026
4.	Population of colonies and dependencies	5,510,000	13,993,000	307,173,631	375,526,224	30,520,293	40,878,248	....	8,277,000
5.	Increase in home population	....	32%	....	19%	....	3%	....	49%
6.	National expenditure	324,869,000	696,100,000	425,679,675	837,017,000	617,151,902	846,587,000	365,773,905	654,137,998
7.	National debt	345,071,000	1,224,808,000	3,282,115,000	3,567,498,000	5,908,658,000	6,280,791,000	915,962,112	1,015,789,000
8.	Production of coal	94,203,200	234,521,200	185,479,126	271,900,000	26,025,000	38,023,000	150,528,713	443,025,000
9.	Production per head	1.94	3.59	4.89	5.99	.67	.96	2.40	4.74
10.	Production of pig iron	4,641,217	15,280,527	7,406,000	9,874,620	1,897,000	4,410,856	8,279,870	24,027,733
11.	Production per head	.09	.23	.19	.22	.05	.11	.13	.26
12.	Cotton spindles	(estimated) 5,000,000	10,300,000	44,750,000	56,500,000	(1887) 5,039,263	7,200,000	14,781,000	29,803,000
13.	Production of wheat	103,800,436	149,411,000	72,127,000	66,289,000	212,480,000	315,126,000	611,780,000	621,338,000
14.	Production of rye	230,531,635	456,600,000	....	1,500,000	59,369,000	50,936,000	33,000,000	35,664,000
15.	Production of potatoes	855,103,710	1,263,024,000	113,978,666	280,753,000	410,396,000	423,573,000	150,000,000	292,737,000
16.	Production of oats	200,306,150	530,764,000	112,386,000	177,170,000	291,899,000	303,328,000	738,394,000	922,298,000
17.	Value of exports	755,769,000	1,928,419,000	1,202,138,000	2,209,972,000	688,177,000	1,172,834,000	884,480,810	2,013,549,000
18.	Exports per head	15.29	29.57	31.73	48.71	17.94	29.61	14.13	21.56
19.	Rate of increase per head	....	93%	....	53%	....	65%	....	53%
20.	Value of imports	1,007,890,000	2,309,756,000	1,816,392,000	3,309,987,000	919,132,000	1,556,705,000	844,916,196	1,527,966,000
21.	Mileage of railway lines	26,627	37,995	20,191	23,350	21,515	30,686	166,832	241,196
22.	Tonnage of merchant marine	1,433,413	4,397,098	8,933,494	19,344,487	786,630	1,958,645	1,936,252	2,998,457

CHAPTER III

The Genesis of Anglo-German Reciprocal Suspicion

Bismarck's foreign policy as Imperial Chancellor. Germany's unassailable position in 1890. Character of William II. "Dropping the Pilot." Dual Alliance formed. Kingship by the Grace of God. German *Welt-politik*. The "Kruger Telegram." East Asiatic Expedition; Kiau-Chau. Naval Bill of 1898. Admiral von Tirpitz, Imperial Secretary for the Navy. German Navy League. Bundesrat incident. Naval Bill of 1900. Isolation of Great Britain. German attitude during the South African War.

Bismarck's practical, unromantic foreign policy as Imperial Chancellor during twenty years was on the whole favorable to European peace. He concentrated his efforts on consolidating the position which Germany had acquired in central Europe, and in this purpose he was eminently successful. He knew that France cherished the hope of revenge and of an opportunity to recover the lost provinces. But he was convinced that France alone was powerless, and for protection against a possible combination of foes, he had forged the powerful Triple Alliance. He saw with satisfaction, moreover, the estrangement of France and Great Britain. He probably helped to make Russia harmless to Germany, and an object of dread to England, by encouraging her to extend her influence in Asia. Bismarck's fixed rules for German foreign policy were to keep in sympathetic touch with St. Petersburg, and never to quarrel with Great Britain. With regard to the position of Germany on the sea he said that if they should build up a navy as strong as England, they should still have to fear an alliance of England and France. Yet during his chancellorship most of the German colonial territory was acquired, and acquired,

DROPPING THE PILOT.

Cartoon by Sir John Tenniel which appeared in *Punch*, March 29, 1890, apropos of the dismissal by William II of Bismarck from the office of chancellor.

moreover, without arousing noteworthy suspicion. The colonial empire was inaugurated by a Bremen merchant in securing a cession of land from the Hottentots in 1883, which became the nucleus of German Southwest Africa. The Kamerun and Togo followed in 1884, and about the same time Dr. Carl Peters founded the German East Africa Company and obtained concessions from the Sultan of Zanzibar. The narrow strip of territory on the coast fronting the present German East Africa was leased by the Sultan of Zanzibar to the Germans for fifty years in 1888; and two years later the Sultan's rights were purchased for about $1,000,000. A German protectorate had been established over what is now called Kaiser Wilhelm's Land, the northern section of south-east New Guinea, with adjacent islands, in 1884.

In 1890 the position of Germany was unassailable. She was easily the strongest military state, and besides, she was supported by her two allies, while France was still alone, and Great Britain pursued her policy of magnificent isolation. But within a very short time France and Russia were linked together as allies, fourteen years later France and Great Britain had reached a cordial understanding, and eighteen years later Great Britain had been joined to France and Russia in a bond of friendship only one degree removed from an alliance. This profound, and decided alteration in the diplomatic aspect of Europe as affecting the position of Germany is a matter of the utmost historical importance. Investigation will be busy for a generation at least in explaining the causes of this transformation. We may venture to make some general observations regarding them a little later. For the present, we must turn our attention to the young ruler who was about to supersede the veteran Chancellor in the guidance of the nation's policy.

The present German Emperor, and King of Prussia, William II, was born January 27, 1859, the son of the late

Emperor Frederick and the Empress Victoria, who was daughter of Queen Victoria. He ascended the throne, June 15, 1888, after the death of his father, who had ruled only ninety-eight days. To his teachers, perhaps, is partly due his solemn conception of the responsibility of his office. The seemingly mystic, mediaeval trait in his character suggests the temperament of his romantic great-uncle Frederick William IV, and contrasts strangely with his well-known zeal for the modern, materialistic interests of life. He possesses a broad, and sympathetic nature. His mind is well informed and intelligent, but probably not exceptionally profound. The versatility of his interests and enthusiasm makes him seem at times erratic. He has been inclined to emphasize the decorative element in government and foreign relations, perhaps shrewdly, although this tendency has frequently been criticized as sensational, or even theatrical. He has displayed independence, although not always profound discernment, in his choice of men and means. His tireless activity has not always permitted him the time to weigh carefully in advance the probable effect of his words and actions.

William II at the age of twenty-nine, when he came to the throne, was impulsive, filled with generous intentions, and impatient of tutelage. The Chancellor was cautious, cynical, and accustomed to exercise authority. Misunderstandings between them were unavoidable. The details of the progress of their estrangement do not concern us. It will suffice to note the immediate cause of their final rupture, since it is significant. Bismarck protested against the unusual conduct of William II, as King of Prussia, in communicating directly with the different ministers on official business. He declared that only the Minister President, not the ordinary ministers, should have audience with the Crown. His claim was based upon a royal order of 1852, and his words may appropriately be quoted, since they represent the position of the two men in an interesting controversy:

"If each individual minister can receive commands from his sovereign without previous arrangement with his colleagues, a coherent policy, for which someone is to be responsible, is an impossibility. It would be impossible for any of the ministers, and especially for the Minister President (of Prussia), to bear the constitutional responsibility for the cabinet as a whole. Such a provision as that contained in the order of 1852 could be dispensed with under the absolute monarchy, and could also be dispensed with to-day if we returned to absolutism without ministerial responsibility. But according to the constitutional arrangements now legally in force the control of the cabinet by a president under the order of 1852 is indispensable." It was clearly a contest for the conception of the cabinet as an independent, united body against the disintegrating influence of monarchical interference with the members individually. We respect the dignified remonstrance of the old Chancellor, calm in the assurance of a well-earned title to authority, in vindication of his conception of ministerial responsibility, even though it is not the variety of responsibility which is embodied in parliamentary government. The world naturally assumed that the Kaiser would establish his direct personal control in the affairs of state, that he would virtually be his own Imperial Chancellor. This belief, together with his picturesque personality, has made the Kaiser the most conspicuous monarch in Europe. Bismarck's resignation as Imperial Chancellor and Prussian Minister President, occurring March 18, 1890, was in this respect a turning point in the history of the modern German Empire. The Kaiser announced this event in the following despatch;—"I am as much afflicted as if I had lost my grandfather anew; but we must endure whatever God sends us, even if we should have to die for it. The post of officer on the quarter-deck of the ship of state has fallen to me; the course remains unchanged. Forward with all steam!"

William II dropped his pilot, took his place on the quarter-deck, and with confidence in his own foresight and the destiny of the Fatherland soon changed the direction of the ship of state to a bolder, less familiar course. Since this memorable event four Chancellors have served under him, their individual character largely effaced beneath the impression of his masterful personality. These have been Caprivi, 1890–94; Hohenlohe, 1894–1900; von Bülow, 1900–09; and Bethmann-Hollweg since July, 1909.

We shall review briefly the course of the reign of William II to the close of the international crisis of 1911, selecting the incidents of greatest significance for our purpose, and repeating as far as possible the words of the principal personalities themselves, the Kaiser, his advisers, and his opponents, in explanation of their views. We may then undertake to formulate our conjectures relative to the causes of the greatly altered conditions in the political world, and an opinion as to the extent to which the policy and activity of Germany involved potential causes of the war.

A short time after the resignation of Bismarck, the so-called Re-insurance Treaty with Russia, which had been negotiated by the great Chancellor, was dropped by Germany. Among the last words of the old Kaiser, William I, on his death-bed, had been an exhortation to maintain friendly relations with Tsar Alexander III, and never to quarrel in that quarter. The action of the German government in renouncing this mutual guarantee of friendship probably strengthened the forces that were drawing Russia and France together. A French naval squadron brought the greetings of the Republic to the Tsar at Cronstadt the following summer, and about this same time a secret treaty

Kaiser William II, wearing the uniform of an admiral

Prince Bernhard von Bülow, Chancellor of the

of alliance between the two countries was probably signed. Its existence was revealed by the words of Tsar Nicholas II, during President Faure's official visit at St. Petersburg in 1897, when he alluded to the "two friendly and allied nations."

Unfavorable comment has often been made on the Kaiser's belief that his authority has come to him "by the Grace of God." A few extracts from his own discourses will serve to illustrate the nature of this conviction. Speaking in Königsberg, the ancient capital of Prussia, the Kaiser said:

"Here my grandfather placed, by his own right, the crown of the Kings of Prussia on his head, once more laying stress upon the fact that it was conferred upon him by the Grace of God alone, not by parliament, by meetings of the people, or by popular decisions; and that he considered himself the chosen instrument of heaven, and as such performed his duties as regent and as ruler. Considering myself as an instrument of the Lord, without being misled by the views and opinions of the day, I go my way, which is devoted solely and alone to the prosperity and peaceful development of our Fatherland."

At the ceremony of unveiling the colossal monument of William I, which rises at the extremity of the point, between the Moselle and Rhine at Coblentz, September 1, 1897, the Kaiser referred to his grandfather in these terms:

"He came forth from Coblentz to ascend the throne as a chosen instrument of the Lord, and as such he regarded himself. For all of us, and particularly for us sovereigns, he again raised on high a precious jewel, and made it sparkle with bright rays, a treasure which I trust we may hold high and sacred. It is kingship by the Grace of God, kingship with its onerous duties, its endless, constant toils and tasks, with its tremendous responsibility before the Creator, from which no mortal, no minister, no chamber of deputies, no nation can release the sovereign." "For me it shall be an exalted duty to walk in paths designated for us by that great ruler, in my solicitude for my country to hold my hand over the glorious jewel to which I have referred, and in accordance with the old tradition which stands firmer than iron or the walls of Ehrenbreitstein, to take this province to my heart and tend it with a father's care."

His expressions relative to the divine right of the kings of the Hohenzollern family were severely criticised by some of the German papers.

On the occasion of the dedication of a monument to the Great Elector in Berlin, the Kaiser, after remarking upon the glorious record of his predecessors, continued as follows:

"How is this wonderful success of the House of Hohenzollern to be explained? Solely in this way, that every prince of the House is conscious from the beginning that he is only an earthly vicegerent, who must give an account of his labor to a higher King and Master, and show that he has been a faithful executor of the high commands laid upon him."

In spite of the Kaiser's emphatic repetition of his conviction of a divine calling, it is possible that too much importance has been ascribed to his claim to rule "by the Grace of God." The formula passes entirely unnoticed when employed with the titles of other monarchs, who do not exert themselves to lay stress upon it. It is somewhat difficult, moreover, to determine precisely the practical meaning which the Kaiser himself would ascribe to the oft-asserted theory. It is a historical fact, no doubt, that the Hohenzollern princes did not obtain their power by popular authority. Since the Kaiser seems to be a very devout man, perhaps to his own consciousness the commission to exercise royal authority by divine command is different in degree only from the duty of every mortal to perform with energy and a deep sense of responsibility the function, to which the will of God, manifesting itself through the circumstances of human environment, has called him. One cannot but be impressed by the deep sense of responsibility which the Kaiser appears to bear in consequence of his belief in the divine sanction of his prerogative.

The practical constitutional significance of the theory of "divine right" for the German Empire and Kingdom of Prussia is restricted to the fact alone that ministers are not responsible to parliaments. If parties were developed strong enough to control compact majorities in the Reichstag and the Prussian Diet, the vitality would fade from this arrogant doctrine at the mere threat of withholding supplies, and it would be reduced to the shadowy state of a meaningless historical formula.

German policy was profoundly modified during the concluding decade of the nineteenth century. We shall not endeavor at present to define the new aims of German diplomacy, except to note that it intended to make its presence felt in all parts of the world. German foreign policy became a *Welt-politik*. The enlargement of its scope, and the intensifying of its activity were due to a variety of causes. The rapid development of German commerce, industry, and capital required room for expansion. The territorial extension of German influence was absurdly restricted in proportion to the intellectual vigor, physical robustness, and military superiority of the nation. The conviction was growing, based upon the manifold national achievements, that the Germans possess superiority in civilization, and are therefore called by destiny to a position of leadership among the nations.

At the same time the emigration of thousands of sturdy, energetic Germans to foreign countries, where they would be lost to the Fatherland, was regretted as a serious national misfortune. Great Britain and France, which had become consolidated by centuries of national unity, had secured a prior claim by occupation or conquest to the most desirable lands for European colonization before Germany solved her most pressing internal problems and could employ her energies abroad. A feeling gained strength in Germany that any opportunity to obtain compensation for these unfavorable conditions ought to be immediately seized. German *Welt-politik* was in part, therefore, the response to a popular demand for a more forcible expression of the national life.

The Kaiser made himself an interpreter and leader of the expansive tendency. On his youthful visits to England, his ardent, imaginative spirit must have been deeply impressed by his grand-mother's title, Empress of India, the imperial range of English political life, and the proud supremacy of the seas. In comparison, German life, doubtless, seemed in those days rather provincial and confined. As Kaiser, William II retained an admiration for all that was excellent in England, and naturally enough he endeavored to reproduce the elements of English superiority in Germany. German imitation of things English has been conspicuous in different ways during the past century. As long as the Germans confined their imitative zeal to the comparatively harmless field of letters or of sports or even to awkward attempts at setting up amongst themselves the constitutional machinery of English parliamentary government, the British public looked on with approval or good-natured condescension. But as soon as this surest form of flattery was extended to the sphere of naval enterprise, the spirit of friendly criticism was suddenly transformed into a feeling of apprehension. The relationship between the two powers became the subject of anxious consideration.

The Kaiser preached the doctrine of the expansion of German enterprise, and as a necessary companion to it, the idea of a big navy. In Berlin in 1896 he said: "The German Empire becomes a world-empire. Everywhere in the farthest parts of the earth live thousands of our fellow-countrymen. German subjects, German knowledge, German industry, cross the ocean. The value of German goods on the seas amounts to thousands of millions of marks. On you, gentlemen, devolves the serious duty of helping me to knit securely this greater German Empire to the Empire at home."

And at Hamburg in 1899, during a noteworthy address, of which the following extracts reveal his cherished ambition, as well as his problems, he said: "A strong German fleet is a thing of which we stand in bitter need." "In Hamburg especially one can understand how necessary is a powerful protection for German interests abroad." "Our people must resolve to make some sacrifice. Above all they must put aside their endeavor to seek the excellent through the ever more sharply contrasted party factions. They must cease to put party above the welfare of the state as a whole. They must curb their ancient and inherited weakness, to subject everything to the most unlicensed criticism; and they must stop at the point where their most vital interests become concerned. For it is precisely these political sins which revenge themselves with such effect on our sea interests and our fleet. If the strengthening of the fleet had not been refused me during the past eight years of my government, in spite of all appeals and warnings—and not without raillery and abuse for my person—how differently could we not have encouraged the extension of our trade and the expansion of our interests beyond the seas."

The same year in Stettin the Kaiser coined the expression which more concisely than any other signifies the new policy;—"Our future lies on the water."

Toward the close of 1895, when conditions in the South

African Republic were approaching a crisis, the Pretoria government sent a message to Berlin inquiring what attitude the German government would assume if hostilities broke out between themselves and the British Empire. Also, a few days before the famous Jameson Raid, the German residents in the Boer capital requested by cablegram to Berlin the protection of a detachment from a German warship that lay in Delagoa Bay, on the coast of Portuguese East Africa. A reply by letter had already been sent to the Transvaal government. Yet the two communications from South Africa help to explain the momentous "Kruger Telegram," which startled the British Empire, and astonished the world.

Dr. Jameson's misguided venture occurred on December 29, 1895, and January 3, the Kaiser repaired to the Chancellor's palace to meet some of his ministers for the discussion of the event. We must not forget that the German Empire in its colony of Southwest Africa was almost a neighbor of the Transvaal. The text of an imperial telegram of congratulation to President Kruger had been prepared at the instance, very likely, of Baron Marschall von Bieberstein, the energetic Secretary for Foreign Affairs, who urged that the upholding of the independence of the South African Republic was in Germany's interest. After some hesitation, and alteration, the Kaiser signed the fateful message, believing, probably, that in its final form it was sufficiently non-committal and inoffensive.

The following is the text of this historic message:

"I congratulate you most sincerely on having succeeded with your people, and without calling on the help of foreign powers, by opposing your own force to an armed band which broke into your country to disturb the peace, in restoring quiet and in maintaining the independence of your country against external attack."

Harmless as well as affable it may seem at the first glance; but we must recall the London Convention of 1884, by which the foreign relations of the South African Republic were committed to Great Britain. In its sudden effect, the "Kruger Telegram" was an electric spark igniting a mine of excitement and emotion. The violence of the explosion reverberated for years. A wave of indignation passed over Great Britain, and the press expressed emphatically the general feeling of resentment. It seemed to the English that the German government assumed the right to ignore treaties between third parties and themselves. Vigorous measures were demanded of the government, which at once responded by an official statement affirming unequivocally the British control of the foreign affairs of the Transvaal. Extraordinary naval activity followed immediately as a demonstration to enforce the British position. Six battle-ships were gotten ready to form a flying squadron, measures were taken for mobilizing a part of the reserve fleet, and a flotilla of torpedo-boat destroyers was brought together. The German government, surprised at the effect of their message, explained within twenty-four hours that no offence was intended, and the inspired press reiterated their declaration.

Technically the German government had committed a diplomatic blunder. The practical effects of the "Krüger Telegram" were deplorable. If the motive which had prompted it was disinterested, the telegram was certainly superfluous, and the risk of misunderstanding gratuitous. If it was part of a design for extending German influence, it was manifestly fruitless, since it was not followed up even by diplomatic action, when the occasion for it was offered. It encouraged the Boers to entertain a delusive hope of German assistance, and it made the German foreign policy appear either vacillating or timid.

On the other hand, the excitement in Great Britain was excessive. The English could logically protest at the sending of a communication from the German government directly to the head of the South African Republic. But the expression of sympathy in itself was no just cause for indignation, since Jameson's Raid had been entirely unjustifiable. Moreover, in view of the very prompt disavowal of any unfriendly intention on the part of the German government, the naval demonstration might well have been discontinued. As it was, the unusual parade of naval power seemed to be intended to overawe and intimidate the Germans. And by impressing upon them a sense of their own weakness on the sea, it was one of the factors that eventually induced them to build up a strong navy. From this time, unfortunately, the relations between Germany and Great Britain have never been free from suspicion.

Certain events of 1897–8 are significant for the light which they throw on the development of German *Weltpolitik*. News of the murder of the two Catholic missionaries in the Chinese province of Shantung was received, November 1, 1897, and orders were promptly despatched to the East Asiatic squadron to proceed at once to Kiauchau Bay, a convenient harbor, lying nearest to where the crimes had been perpetrated, for the purpose of demanding reparation. On November 23, Prince Henry, the Kaiser's brother, was made commander of a second squadron, to be sent from Germany to the support of the first. In view of the Kaiser's enthusiasm for naval affairs, we are not surprised that he made the departure of the first naval expedition undertaken by the German Empire an occasion of great solemnity.

In a farewell banquet to Prince Henry at Kiel, December 15, the Kaiser, after alluding to the former maritime glory of the Hanseatic League, remarked that the Hansa decayed,

FREDERICK WILLIAM

Crown Prince of the German Empire.
In field uniform.

PRINCE HENRY OF PRUSSIA

Wearing the uniform of a German admiral.

"and could not but decay, for the one condition, imperial protection, was wanting. Now things are different. As the first preliminary condition the German Empire has been created. As the second preliminary condition, German commerce is flourishing and developing; and it can develop and prosper securely only if it feels safe under the power of the empire. *Imperial power means naval power*, and they are so mutually dependent that the one cannot exist without the other." As explaining the spirit of the instructions to his brother, he continued; "May it be clear to every European foreigner out there, to the German merchant, and above all, to the foreigner whose soil we may be on, and with whom we shall have to deal, that the German Michael has planted his shield adorned with the eagle of the Empire firmly on that soil, in order once for all to afford protection to those who apply to him for it." "Should, however, any one attempt to affront us or infringe our good rights, then strike out with the *mailed fist*, and if God will, weave round your young brow the laurel which nobody in the whole German Empire will begrudge you." In the course of his reply Prince Henry employed the following expression; "I am not allured by hopes of winning glory or laurels; I am only animated by one desire—to proclaim and preach abroad to all who will hear, as well as to those who will not, the gospel of your Majesty's anointed person."

Derision of the romantic tone of this ceremony, more than opposition to the undertaking itself, was the chief feature of the observations of the leading British journals. It is noteworthy that intelligence of the arrival of a Russian fleet at Port Arthur, with evident intention of making it a permanent winter port, was received four days after the farewell to Prince Henry. The German expedition obtained a lease for ninety-nine years of the Chinese sovereign rights over a territory of about two hundred square miles, including the ports, Kiau-Chau, and Tsing-Tau, together with an exclusive concession for the construction of a railway to connect these ports with the great Chinese railway system. It was equivalent to a German economic protectorate over the populous province of Shantung, with the actual possession of points of tremendous commercially strategic importance.

The Secretary for Foreign Affairs, von Bülow, addressed the Reichstag, February 8, 1898, relative to the expedition to China, as follows: "The expedition of the cruiser squadron to Kiau-Chau was not the result of a sudden impulse but of a decision reached through careful deliberation of all the circumstances. It is the expression of a prudent, consistent policy. We were already fully alive to the necessity of a territorial foothold in East Asia, without which our economic, maritime, and political interests would be deprived of the security afforded by a definite point of coördination. In an economic connection we require such a door to the Chinese exploitive area as France possesses in Tongking, England in Hongkong, and Russia in the north. The Chinese Empire with its population of nearly 400,000,000 is one of the world's richest markets of the future. We cannot allow ourselves to be excluded from this market, upon which our economic and material, and consequently political and moral, progress depends."

The words of the Kaiser in bidding farewell to his brother, especially the allusion to imperial protection as a necessary condition for commercial prosperity, were the reflection of a proposed fundamental change in the financial policy of the government regarding the naval establishment, a bill for which had already been submitted to the consideration of the Reichstag. The adoption of this proposal, as embodied in the famous Navy Law of 1898, was a turning-point in the reign of William II. This innovation was to sanction in advance a comprehensive program of naval construction covering several years, together with the necessary current expenses of the fleet, instead of voting yearly appropriations as heretofore. It extended to the navy the system according to which the funds were granted for the army, formerly for periods of seven years, but at that time for terms of five years' duration. Thus the Reichstag was to pledge in anticipation the annual contributions required for the fleet, and in return the government bound itself not to increase within the specified time the measure of its requirements. In introducing these proposals on December 6, the Chancellor, Prince Hohenlohe, remarked that under the old system disputes often arose over appropriations for a cruiser or two which in their violence were out of all proportion to the object in view. "The result of this method of handling naval questions has been," he said, "that in spite of considerable appropriations our fleet has not developed in such a way as the vital interests of the country require. The navy has had to live from hand to mouth, so to speak, and it has been impossible to follow a consistent policy in its development. We have, accordingly, become convinced that the system is faulty. A substantial building cannot be erected unless the builders have a definite plan, according to which the foundations are laid, and the walls are reared. The proposal which has been submitted to your deliberation is calculated, in the same way, to unify all the coöperating factors." He went on to say that while the measure placed a limitation for some years on the freedom of the Reichstag in voting supplies, it supposedly tied the hands of the government also. With reference to Germany's naval intentions, he remarked that Germany had no idea of vying with the great maritime powers, but that if her interests were affected she must be able to speak plain German through her navy.

At that time the German navy consisted of seven first, five third, and eight fourth class ironclads; four first, six second, three third, and eight fourth class cruisers; thirteen armored gunboats; and smaller craft. It was proposed to reduce the different classes of larger war-vessels to three, namely, battleships, first-class cruisers, and second-class cruisers; and the bill called for seven battleships, and nine cruisers, to be constructed within seven years at an aggregate cost of about $105,000,000. This non-recurring expenditure, together with the usual, or recurring expenses, was to be distributed over the period in such a way that the total appropriation for naval purposes each year would increase gradually from $29,208,000 for the fiscal year 1898–9 to $35,928,000 for 1904–5. The large cruisers to be provided were a rather unique class at that time, having a tonnage of 8860, engines of 15,000 horse-power, triple screws, and a high speed, about twenty-one knots.

On the occasion of the discussion of the naval proposals the Foreign Secretary, Count von Bülow, cited the oriental situation as illustrating the intention of the government to maintain a firm, but not aggressive, foreign policy, and employed an oft quoted expression. "We are animated," he said, "with the most benevolent and friendly intentions regarding China, and have no wish either to offend or provoke her." "It is our duty to insist that the German missionary, the German merchant, German goods, the German flag, and German shipping enjoy the same respect in China as those of other powers. We cheerfully observe the same consideration for the interests of other great powers in East Asia, which we expect that they will display towards ours. In short, we desire to put no one in the shade, *but we too demand our place in the sun.* We shall endeavor in accordance with the traditions of German policy, without unnecessary rigor, but also without weakness, to guard our rights and our interests."

During a subsequent debate on the Navy Bill he said

Theobald von Bethmann-Hollweg, Chancellor of the German Empire and President of the Prussian Ministry.

that Germany had to provide against being excluded from competition in promising countries, and that the day was past when the German was content to resign the land to one power, the sea to another, and to allot to himself only the heavens, the abode of pure and abstract doctrine. The Navy Bill passed the Reichstag, March 28, 1898.

Late in the same year Germany further strengthened her position in the Orient by purchasing from Spain the Caroline and the Ladrone Islands, with the exception of Guam, for a sum equivalent to about $4,000,000.

The Kaiser was the prophet, the apostle, and might even be called the high-priest of the modern German navy; for the project of a powerful navy was proclaimed and cherished with the fervent devotion of a religious cult. But the maker of the navy, if the title can be conferred upon any individual, was Chief Admiral Alfred von Tirpitz, and his commanding personality has been inseparably connected with the development of German sea-power from the time when he was appointed Minister of Marine in 1897 until to-day. An eloquent commentary upon his association with German naval policy is the growth of the annual appropriations for naval purposes, while he has been minister, from about $30,000,000 in 1898 to $116,700,000 in 1913. His comprehensive experience in naval affairs, acquired in the course of a career extending from cadet through various grades, in an orderly line of promotion, to minister, is a foundation for unwavering assurance in his authority. He is a character of indomitable energy and determination. As chief-of-staff at the Kiel naval station he displayed conspicuous executive ability by introducing into naval administration the standard of efficiency which prevailed in the army. As minister he exhibits a rare combination of practical seamanship, administrative talent, and discerning statesmanship.

The diplomatic shrewdness of this great naval minister has been repeatedly demonstrated in the cleverness and tact with which the government, under his leadership, has employed every available means of exerting influence on the public for creating a favorable atmosphere just before inaugurating each successive scheme for the enlargement of the fleet. The Navy League has been an effective auxiliary in this propaganda. It is a voluntary association which undertook to spread the gospel of German sea-power. It mobilizes the forces of public opinion favorable to a vigorous naval policy. It has branches in all parts of the empire, publishes a periodical, sends out lecturers, arranges for excursions of school-children to visit units of the fleet, and disseminates its message to the people in other practical ways.

The South African War had a most unhappy influence upon Anglo-German relations. A short time after the beginning of hostilities a British cruiser stopped a German mail-packet steamer, the *Bundesrat*, off the east coast of Africa, because she was suspected of carrying contraband. Although the *Bundesrat* was quickly released, the incident excited general indignation in Germany. The government seized upon the feeling of exasperation as an opportunity of securing popular support for a vastly more extensive naval program. In spite of the agreement covering seven years implied in the Naval Law of 1898, passed about eighteen months before, the Chancellor complacently informed the Reichstag that a bill was being prepared to amend the earlier measure. The ensuing Naval Law of 1900 may be considered the real beginning of the momentous naval rivalry of Germany and Great Britain.

The government regarded the further expansion of their sea forces as indispensable in consequence of the far-reaching changes in the international situation recently brought about, particularly by the Spanish-American, and South

African, Wars. The sudden disintegration of the Spanish colonial power was an object lesson of sudden revolutions that might be expected, when the vigorous nations with strong navies at hand would derive profit. Portugal, for example, was even weaker than Spain. In fact, her empire seemed to be on the point of collapsing from bankruptcy. Her population was about 5,300,000, and her public debt about $600,000,000. Her dependencies embraced an area of 801,060 square miles with a population of 9,216,707, and the largest part, Portuguese East Africa, was admirably situated for becoming an addition to the German colonial empire.

The Foreign Secretary, Count von Bülow, in a rambling speech before the Reichstag, made the following remarks, which fairly represent the government's attitude: "Germany cannot stand aside while other nations divide the world among them. The rapid increase of our population, the growth of our industry, the capacity of our merchants, in short, the keen vitality of the German people, have drawn us into the international market and bound our interests up with those of the whole world."

But the impression was immediately produced in some quarters that the naval policy of the German government was directed against Great Britain, with the hope of ultimately wresting from that power the supremacy of the seas, and henceforth every fresh proposal for increasing the navy has encountered this suspicion in the Reichstag. Is it any wonder that what many Germans suspected, many English believed? A speech of Herr Richter of the radical opposition in the Reichstag expressed the view of those who denounced an attitude of hostility against England:

"It is not true," he said, "that England is hostile to us in our colonial aims. England could have taken all our colonies long ago, if she had thought it worth her while, for they all lay at her door. It has been possible for all our

Imperial Chancellors from Prince Bismarck downwards to delimit our colonial spheres of interest with England in a business-like manner. Who would ever have imagined that England would have ceded Heligoland (in 1890) to us? All these agreements were successfully concluded without any regard to our navy, but as a result of the general attitude of Germany to England."

Admiral Tirpitz introduced the bill, February 8, 1900. It embraced a scheme of constructive activity continuing until 1920, and providing for two double squadrons of seventeen battle-ships each, fourteen cruisers of the first class, thirty-eight smaller cruisers, and ninety-six torpedo boats and destroyers. The aggregate sum which it was necessary to raise to 1920 for non-recurrent expenses was about $200,000,000. In the course of the debate, Herr Bebel, the veteran chieftain of Social Democracy, affirmed that it was evidently the intention of the government to create a fleet capable of coping with that of England, and that a conflict with that country would be the greatest calamity that could befall Germany; for France and Russia would wait until Germany was exhausted and then fall on her. England and Germany, in his opinion, were natural allies. Von Tirpitz denied, however, that the increase of the navy had necessarily any connection with a conflict with England; it was imposed upon them by general considerations of policy. The preamble of the bill itself sets forth as motive the proposition that "Germany must possess a battle-fleet so strong, that a war with her would even for the greatest naval power be accompanied with such dangers as would render that power's position doubtful." The opinion of von Tirpitz himself has been that the general position of Germany required her to maintain a naval establishment of two-thirds the British strength. The bill was finally passed, June 12. If the flying squadron was England's answer to

The Kaiser. Queen Mary. King George. King Alphonso. The Kaiserin.

Edward VII. The Tsarina. Queen Alexandra. Amelia, Queen of Portugal.

Group showing the close relation of the dynasties of Europe brought about by the intermarriage of royalties.

Germany's blunder in the "Kruger Telegram," the Navy Bill of 1900 was Germany's reply to England's mistake in stopping the "Bundesrath." The race of armaments had begun.

Count von Bülow, who had been Foreign Secretary since 1896, was appointed Imperial Chancellor to succeed Prince Hohenlohe, October 18, 1900. His tenure of this office continued until July 14, 1909, longer than that of any other Chancellor since Bismarck. These were memorable years for Germany in a variety of ways, and von Bülow stands out as the most conspicuous personality among the successors of the Iron Chancellor. He is the embodiment of affability and suavity of manner. He has none of the narrow spirit of the Prussian aristocratic caste. He was distinguished by his skill in debate before the Reichstag. While free from an unreasonable spirit of Chauvinism, he was ready at all times to uphold Germany's right to "a place in the sun."

At the time of the South African struggle Great Britain stood alone amongst the nations. France harbored a feeling of resentment for the continued British occupation of Egypt, and more especially for the humiliation of the recent Fashoda incident of 1898, when England mobilized a flying squadron and imposed her own view upon the French in a territorial question touching the upper valley of the Nile. A traditional rivalry and suspicion separated Great Britain from Russia; and mistrust and prejudice were daily broadening the spiritual gulf between Great Britain and Germany. This was the state of affairs in 1900. Fourteen years later we find Great Britain leagued with six other states. This transformation from isolation to far-extending alliances is one of the most astonishing diplomatic developments in modern history. The change was not planned deliberately. For the Triple Entente between Great Britain, France, and

Russia was not the work of a single statesman, nor the result of a single clearly-defined project. It developed step by step under the influence of circumstances and a broader view of national advantage. It will be instructive to observe that the way began to be smoothed for the first step in these national reconciliations with Great Britain not very long after the great German Naval Law of 1900.

The tone of the continental press was generally unfriendly to Great Britain during the South African War. It was possibly not more bitter in Germany than in France. But the virulent anti-British attitude of most of the German papers was more irritating in its effect, because Germany had sent the "Kruger Telegram," because Germany was supposed to entertain sentiments of envy and jealousy against Great Britain, and because Germany had less cause for resentment than France. During all this time the official attitude of the German government was commendably correct. When President Kruger with his associates visited Europe in the hope of obtaining a diplomatic intervention of the powers, he was received by the president and foreign minister of France. But a telegram from Potsdam informed him that the Kaiser was not in a position to receive him unless he were presented by the British ambassador. The Kaiser's position in the face of the passionate indignation of a large part of the people was a difficult one, as he himself afterwards declared. There were some who insinuated that the government maintained a scrupulously neutral attitude on account of the dynastic connection with Great Britain.

The irritation on both sides of the North Sea was intensified by finding expression in the public utterances of prominent political personalities. Thus Mr. Joseph Chamberlain, Colonial Secretary, in a speech in Edinburgh, October 25, 1901, after announcing that severer measures might be necessary against the guerilla bands, which continued an obstinate resistance prolonging the war to no purpose, went on to say that the British government could find precedents for anything it might do in the action of the nations who criticized the British "barbarity" and "cruelty," but whose example in Poland, in the Caucasus, in Algeria, in Tongking, in Bosnia, in the Franco-German War they themselves had never even approached. Singularly enough, the reference to a German example was the only one of these which seems to have been seriously resented; but German Anglophobia broke all bounds in consequence of the allusion to the measures taken in the Franco-German War. Indignation meetings of the intellectual classes were held all over Germany to protest against the insolent comparison of the national German army, their own fathers and brothers, with the ruffian mercenaries of the British army. It was immaterial that among these "hired ruffians" there were about 60,000 colonials, including the cream of the youth of Canada, not to mention the yeomanry and volunteers of the better classes from the Mother Country.

If any unfavorable comment is ever made on the conduct of the German army, an important element in Germany immediately rallied in anger to its defense, without first inquiring whether the reproach is justifiable, and this principle of the army right or wrong appears to be inculcated in the schools. It is strangely at variance with the attitude of calm detachment of German scholarship in scientific research and abstract speculation. A nation that consecrates any one of its institutions by an intolerant, unreasoning devotion becomes excessively sensitive to the inconsiderate thrusts of those who are strangers to the impressive quality of the object of veneration. The element in German society which cherishes this irrational attitude is more impatient of criticism directed against the army than against the nation itself; and yet they insist that the army is the nation. This shows that to them the institution itself possesses a character of sanctity which it is impious to violate by word or thought.

In France the Socialists and Radicals laid profane hands on the military cult. It received its death-blow in connection with the Dreyfus Affair. Yet French soldiers, and what is perhaps more significant, French officers, are just as spirited in defending liberty and their country as when impelled by mystic devotion to a traditional fetish, upon whose altar independence of judgment and criticism must be sacrificed.

The German government maintained, outwardly at least, an attitude of moderation during this crisis of popular emotion. The semi-official press admitted, however, that Chamberlain's words had in some measure justified indignation, and when the Reichstag re-assembled, January 8, 1902, Imperial Chancellor von Bülow referred to the prevailing state of excitement in the following terms:

"It was altogether intelligible, that in a nation which is so closely bound up with its glorious army as is the German people, the general feeling rose high against the attempt, and even against the appearance of an attempt, to misrepresent the heroic character and the moral basis of our struggle for national unity. The German army, however, stands far too high, and its escutcheon is far too clean, for it to be affected by distorted judgments. With regard to anything of that kind, the remark of Frederick the Great holds good, when he said, on being told that some one had attacked him and the Prussian army: 'Let the man alone and don't excite yourselves, he is biting at granite'." In the course of the debate Herr Bebel said that he failed to understand the outcry against Chamberlain, while Herr Liebermann von Sonnenberg exclaimed that Chamberlain was "the most accursed scoundrel on God's earth," and said that "Germany's veteran soldiers must be protected against comparison with gangs of robbers and packs of thieves; for it is evident that the greater part of the British army is composed of such elements." The cordial relations between the Kaiser and the English royal family were not interrupted at this time. On the contrary, the Prince of Wales, now King George V, visited Berlin for the celebration of the Kaiser's birth-day, January 26, 1902, eighteen days after the above-mentioned debate.

The Germans affirmed throughout the war that their sympathy for the Boers was disinterested; that it was the natural result of admiration for a small people struggling courageously for their liberty, and of national affinity, since the Boers, as Dutch colonists, were an offshoot of Low German stock. These protestations were the expression of a sincere conviction, but a thoughtful analysis might have brought to light less conscious, but more fundamental, causes. An accession to the British Empire in any part of the world would have been a source of bitterness at that time, and this is quite intelligible in view of the hopelessly disproportionate territorial extension of the two powers. Moreover, both the German government and people had doubtless allowed themselves to speculate vaguely upon advantages which might eventually be secured from friendly coöperation with the Boers. A Teutonic South Africa, under German economic hegemony, if not political leadership, had been a shadowy possibility of the future. The British annexation of the two Boer republics crushed these hopes unceremoniously. It seemed to consolidate the British power, and perpetuate the separation of the German spheres of control in East and West Africa. If the Germans had been deeply moved by a disinterested spirit of sympathy for the Boers themselves, they would have rejoiced at the generosity displayed towards them by the victors, and the subsequent formation of the Union, in which the Afrikanders of Dutch descent are the strongest element. But there has been little indication of such a sentiment. On the contrary, the German press showed annoyance at the conclusion of hostilities, and has been alert ever since to observe any signs that might indicate an unsatisfactory condition of affairs in South Africa.

CHAPTER IV

Conditions Fraught with Danger

Germany and Russia in 1904. Agreement between Great Britain and France, April 8, 1904, respecting Morocco. German press. Kaiser's speech in Bremen, March 23, 1905. Russian disasters destroy balance of forces in Europe. Morocco and French policy. Kaiser's visit in Tangier and speech. Germany's attitude. Crisis in Morocco Controversy; France yields; M. Delcassé resigns. Algeciras Conference, 1906. Significance of Germany's Moroccan policy. The "open door." "*Matin* disclosure." The first "Dreadnought." German Navy Law of 1906. Race in armaments. British naval panic in 1909. Anglo-Russian reconciliation and *Triple Entente*. "*Daily Telegraph* Interview," October 28, 1908. Franco-German agreement of February 9, 1909. French occupation of Fez. *Panther* sent to Agadir. Proposal for "compensations." Speech of Mr. Lloyd George, July 21, 1911. Germany accepts a compromise. Disappointment and resentment in Germany.

Germany had never given up the policy of cultivating quite friendly relations with Russia, and the increasing cordiality between England and France probably suggested renewed efforts to secure an intimate understanding with the Tsar, as a sort of safeguard. We must recall that Russia was at this critical period engaged in her struggle with Japan. But during 1904 the military situation did not seem to preclude the possibility of a favorable result for her armies, and the revolutionary agitation had not yet taken on the character of a national upheaval. The friendship of Russia was still desirable; and Germany's neutrality in the war was conspicuously benevolent as far as her great neighbor was concerned. Complaints were frequently made in Germany at that time that the government's attitude toward Russia was unworthily subservient. Herr Bebel with characteristic frankness denounced Germany's policy as one of blacking Russia's boots. Russian deserters escaping across the frontier were promptly arrested and sent back. Russian subjects suspected of revolutionary activity against the Tsar were expelled from Germany. Not only did the Prussian police coöperate with Russian officials who were attached to the Russian Embassy in Berlin for the purpose of observing their fellow subjects in Germany, and who exercised an extraordinary liberty of search, but even nine Germans were brought to trial in Königsberg for conspiracy, sedition, and lèse-majesté against the Tsar!

There were convincing indications that an intimate understanding existed between the two governments. In the first place, the Russians withdrew most of their regiments from Poland, which would have been impossible if the Germans had not guaranteed the security of the frontier, a pledge which probably included the promise of coöperation of some sort in the event of a Polish uprising. Then, on July 28, a new commercial treaty was signed, which was very favorable to Germany. For, although the higher scale of duties on grain, which had been lately introduced in Germany to favor the agrarians, was incorporated in the new agreement, Russia granted very favorable treatment to imports of German manufactured articles. Under this agreement German trade with Russia prospered amazingly. Moreover, an old arrangement was revived in October, which had existed during several generations between Prussia (later Germany) and Russia, by which the military attaché of each power held the rank of military plenipotentiary at the court of the other, with personal access to the sovereign to whom he was accredited. This institution set up a channel of communication parallel with, but independent of, the customary diplomatic connection. It may have put the monarchs into closer personal touch, but probably involved

Georges Eugène Benjamin Clemenceau, founder of *L'Aurore*, in which was published Zola's famous letter regarding the Dreyfus case.

Théophile Delcassé, Minister of Foreign Affairs of the French Republic.

the danger of confusing the orderly progress of negotiation.

The events of 1904 and 1905 deserve our closest attention. These were years when some decisive factors in international policies made their presence felt. Currents in diplomacy, which had been somewhat uncertain until then, took the course to which they have steadfastly adhered until the present time. The most important of these was the tendency for Great Britain and France to act together.

Directly after the Fashoda Incident in 1898, when Anglo-French relations were very strained, M. Cambon was sent to London as representative of the French government with the hope of bringing about a better understanding between the two nations. The process of reconciliation proceeded very slowly at first, but about the time of the termination of the South African War relations began to improve very rapidly. The two nations suddenly became convinced that no insurmountable difficulties prevented the establishment of harmony between them. One by one the outstanding questions were readily settled. A convention signed April 8, 1904, was the final act of agreement consummating the *Entente Cordiale*, or sincere understanding, between the two countries. This agreement was a recognition by each nation of the other's position in Egypt and Morocco respectively. France was to have a free hand in Morocco, England in Egypt. The provisions regarding Morocco authorized France to take measures to tranquillize the country, and to assist the Sultan in introducing the necessary financial, economic, and military reforms. Furthermore, there were to be equal commercial opportunities for all, as in Egypt.

An agreement was made by France and Spain in October, 1904, relative to Morocco, on about the same basis, with a general understanding that French interests took precedence in about four-fifths of Morocco, and Spanish interests in the other one-fifth.

One of the natural results of the understanding with France was the geographical reapportionment of the units of the British fleet, involving the withdrawal of a large part of the naval strength from the Mediterranean Sea, and a corresponding increase in the North Sea.

Mr. Arthur Lee, Civil Lord of the Admiralty, in a speech at Eastleigh, February 2, 1905, said that there had been a complete redistribution of the British fleet, in order to be prepared for possible enemies. The balance and center of naval power in Europe had been shifted during the last few years. They had not so much to keep their eyes upon France and the Mediterranean as they had to look with more anxiety, though not with fear, towards the North Sea. It was for that reason that the fleets had been distributed to enable them to deal with any danger in that direction. He concluded this analysis of the naval situation by saying: "If war should unhappily be declared, under existing conditions the British navy would get its blow in first, before the other side had time even to read in the papers that war had been declared." Not unnaturally, the Berlin papers looked upon the tone of these remarks as menacing, although it is more than doubtful whether, if the likelihood of such a sudden act of aggression were really contemplated, it would be thus openly proclaimed. The inspired *North German Gazette*, in an article appearing about this time, denied that there was a state of tension in the relations between the two countries, or cause for any.

In considering events of comparatively recent occurrence which still loom large on the horizon of the memory, the zest of curiosity at renewing our first sensations, grown a little dim with the passing of time, impels us frequently to consult the most tangible immediate record of public impressions, the daily press. Whoever would employ the contents of this store-house of contemporary historical evidence for Germany, must use care and discrimination, in consequence of conditions which may be briefly explained.

The tenacity of the intellectual importance of the old local capitals in Germany has retarded the concentration of journalism in a great metropolitan press representing broadly the principal currents of national opinion, such as we find in London and Paris. Some of the other large cities have papers which are no less prominent than the leading journals of Berlin itself. Moreover, the government recognizes the great value of the press as a medium for controlling popular sentiment, and for this reason brings influence to bear upon it in different ways. Some papers, as the *North German Gazette* (*Norddeutsche Allgemeine Zeitung*) reflect very faithfully the attitude of the government, being virtually official organs of publicity. There are other papers, such as the *Cologne Gazette* (*Köelnische Zeitung*), which are semi-official; experimentally inspired, as one might say. The official origin of the government's views, when expressed in these, can be denied, if it is unfavorably received. *The Lokal Anzeiger*, *Berliner Tageblatt*, and *Morgenpost*, are modern Berlin papers with an enormous circulation, but limited political influence. The *Frankfurt Gazette* (*Frankfurter Zeitung*) is a leading Radical organ, and the *Vorwärts* in Berlin, like the Social Democrat party, of which it is the official mouth-piece, represents with uncompromising steadfastness a partisan attitude in opposition to the government. As for a large number of the smaller local sheets, the economic advantage of official advertising keeps them quite subservient to the government's influence. The German papers, on the whole, therefore, must be consulted with discernment, whether one wishes to discover the official attitude, or gauge the popular opinion. The interpretation of leading articles in the semi-official papers presents a baffling, but, perhaps, at the same time enticing, problem for the scholar of recent history.

Events of momentous consequence will directly suggest references to the press to test the responsive beating of the public pulse.

The Kaiser assisted at the dedication of a monument to his father in Bremen, March 23, 1905, and delivered an address which is noteworthy for its apparent ambiguity and the sensational proceeding of which it seems to have served as a prologue. The most striking, and at the same time sharply contrasted, parts of his speech were the following:

"The lessons of history have impelled me to vow never to strive for world dominion. For what has become of the world-empires? Alexander I, Napoleon I, and all the great conquerors swam in blood, and left subjugated peoples, who grasped the first opportunity to assert their independence and shatter the empires of force. The world dominion of which I have cherished the dream consists in this; that the recently amalgamated German Empire should enjoy on all sides the most profound confidence as a tranquil, upright, peace-loving neighbor; and that if future history should speak of a German or Hohenzollern world dominion, it would be founded, not on a policy of the sword, but on the reciprocal confidence of the nations in their struggle towards a common goal; in short, as a great poet has said, 'confined without, but unrestricted in its capacity for development within'."

"Our military preparation on land had fulfilled its necessary development; the time had come for naval armaments to advance. I thank God, that there is no need for the cry of alarm which I once uttered in Hamburg. Our fleet is already on the water, a reality, and naval construction is still proceeding. We have an abundance of seamen; and the zeal and spirit are displayed which characterized the Prussian officers at Hohenfriedberg, Königgrätz (Sadowa),

mperor William II after landing at Tangier, March 31, 1905, on his way to the German Legation, escorted by the uncle of the Sultan of Morocco.

Alphonso XIII, King of Spain, and Prime Minister Eduardo Dato.

and Sedan. And every German ship that is launched is an added guarantee of peace on earth, since it discourages our enemies and raises us in the estimation of our allies."

The prospect of a German Empire dedicated to peace and actively engaged in launching war-ships to ensure the general repose was thus solemnly announced to the astonished nations. The same speech contained a further passage which did not pass without comment:

"We are the salt of the earth; and we must prove ourselves worthy of our high calling. Our young generation must, therefore, learn to deny itself all that is not good for it, to shun the contagion of evil that steals in from foreign lands, and to preserve good morals, discipline, order, reverence, and religion. Then can the German people merit the device inscribed on the helmets of my First Life Guards—*Semper talis*—Always the same."

The world was not left long in doubt as to the significance of this rather incoherent speech, which seemed to be at the same time an attempt to overawe and to reassure. For the Kaiser sailed from Cuxhaven the same day, on the mail steamship *Hamburg*, for a journey to the Mediterranean with Tangier, the commercial metropolis of Morocco, included in the proposed itinerary.

For nearly a year after the conclusion of the Franco-English agreement the German government had shown no signs of solicitude respecting the results of this convention for the affairs of Morocco. The French Foreign Minister M. Delcassé had informed the German ambassador in Paris that the agreement touching Morocco recognized the integrity and independence of the country, and the sovereignty of the Sultan, and authorized the restoration of order under the friendly direction of France. Chancellor von Bülow had declared in the Reichstag, April 12, 1904, four days after the agreement had been concluded, that German interests in Morocco seemed to be in no danger. About the same time, in an interview with King Alfonso of Spain at Vigo, the Kaiser had declared that Germany desired no acquisition of territory in Morocco.

In the mean time, however, disaster after disaster had befallen the Russians in the war with Japan in the extreme Orient. Port Arthur had fallen, January 1, 1905, the Battle of Mukden, March 6–10, had been an irreparable defeat for the Russian army, and the last hope of the Russian sea-power had been shattered in the naval battle in the Straits of Tsushima, May 27. It was commonly believed that Russia had received a blow from which her military power could scarcely recover within twenty years. The equilibrium of military power seemed to have been subverted. The powers of the Dual and Triple Alliances had been facing each other for a number of years. The pressure of one group neutralized the opposing pressure of the other, and thus a balance of forces was effected, and consequently stability. But when the pressure on one side was partly removed, the forces on the other almost inevitably expanded. The expansive vigor of German diplomacy quickly made itself felt in connection with Morocco.

Morocco occupies an area of about 219,000 square miles, and has a population of about 5,000,000 souls. Its soil is fertile in parts, and conceals valuable mineral deposits, notably iron-ore. If one considers first the advantageous situation of Morocco, fronting on the Mediterranean Sea and Atlantic Ocean and flanking the Straits of Gibraltar, and then the character of its native inhabitants, who are entirely incapable of developing the many opportunities which they have at hand, it appears extraordinary that the country was not brought under the controlling influence of any European power before the twentieth century. Naturally France, whose important colony Algeria bounds Morocco on the east for a long distance, had the greatest interest in the establishment of peace and prosperity in the country. The Sultan exercised a precarious authority over the turbulent tribesmen to whom the greater part of Morocco was abandoned. The state of anarchy in Morocco was a subject of grave consideration for the rulers of Algeria.

The natural consequence of the Anglo-French agreement of 1904 was the decision of the French government to press upon the Sultan the necessity of an improvement in the internal situation of Morocco, which was to be undertaken with their friendly assistance. Accordingly, in January, 1905, a special French minister was sent to Fez, the capital of the country, to lay before the Sultan a program of the reforms that were considered necessary.

The Sultan hesitated to accept these proposals, and the French minister, to overcome his reluctance, was reported as having declared that he was acting in this negotiation as practically the representative of all Europe. As early as July, 1904, a loan for the Sultan had been raised in Paris, in consequence, undoubtedly, of the Anglo-French agreement. At that time the Sultan must have known the provisions of the agreement regarding Morocco and must have accepted them, at least tacitly. His opposition to the proposals of the French government in 1905 can best be explained on the assumption that he was encouraged to resist by the German diplomatic representative in Fez. The Sultan even demanded information of the German envoy respecting the character of the French minister's professed commission as representative of Europe, and received the information that Germany, for her part, had not given her authority for such a mandate.

On March 29, a few days after the Kaiser's departure for the Mediterranean, Chancellor von Bülow addressed the Reichstag on Germany's position with regard to Morocco, employing during his speech the following expressions:

"We have in Morocco, as in China, a great interest in the maintenance of the 'open door'; that is to say, equal opportunities for all commercial nations. The German interests are considerable in Morocco; and we must see to it that they have equal treatment with those of other nations." "While the attitude of the diplomat must adjust itself to individual circumstances, the general attitude of German policy has not changed." "If any attempt should be made to change the international status of Morocco, or restrict the 'open door', we must with greater watchfulness than heretofore guard our economic interests from harm."

Two days later, the Kaiser addressed the German colonists in Tangier in a speech which created a world-wide sensation, since it was undoubtedly intended as an emphatic, official statement of the attitude of the German government. He said:

"I rejoice to make acquaintance with the pioneers of Germany in Morocco, and to be able to say to them that they have done their duty. Germany has great commercial interests there. I shall promote and protect our trade, which shows a gratifying development, and make it my care to secure full equality with all nations. This is only possible when the sovereignty of the Sultan and the independence of the country are preserved. Both are for Germany beyond question, and for that I am ready at all times to answer. I think my visit to Tangier announces this clearly and emphatically, and will doubtless produce the conviction that whatever Germany undertakes in Morocco will be negotiated exclusively with the Sultan."

The Kaiser's visit in Tangier was a very drastic assertion that Germany recognized no influence in Morocco obscuring the Sultan's full sovereignty. The existence of a Moroccan Question had scarcely become known to the world at large,

Delegates at the conference held at Algeciras in February, 1906, to consider the claims of France and Germany in regard to Morocco.

when it passed into the acute stage. The Kaiser's visit was like a Kruger Telegram rendered many times more emphatic. But there was no diplomatic blunder this time. The German government had assumed a logically defensible position. They were prepared to back argument by force. They won their point, scoring thereby a diplomatic victory, which was considered a serious humiliation for France. The German view was based upon an international agreement. At a conference of powers in Madrid for the consideration of Moroccan affairs, a convention had been signed July 3, 1880, declaring the independence of the country, and guaranteeing the treatment of the most favored nation in Morocco to all the signatories of the agreement. Germany had been one of the powers which participated in the Madrid Convention in 1880. Germany took the logical view that France could not assume a special position in Morocco tantamount to a protectorate without the consent of all the powers which had signed that agreement. The German papers spoke of the impending "Tunisification" of Morocco. On the other hand, the impression was current in France that German interests in Morocco were so small, about one-tenth of the foreign commerce, that Germany's intervention was an inexcusable act of provocation.

The *North German Gazette* observed on April 5 that the Kaiser's words were in harmony with the statement of the Chancellor in the Reichstag, March 29, that Germany desired equality of treatment in an economic sense for all powers in Morocco. It went on to affirm, however, that France demanded, according to M. Delcassé's own words, an exceptional position in that country, which was apparently inconsistent with the principle of the "open door." The French representative at Fez had assumed the character of mandatory of all Europe, but Germany for her part had conferred no such mandate. Other German papers expressed the opinion that German interests had not suffered injury up to that time, and hoped that Morocco would not undergo the fate of Tunis, from which foreign enterprise had been almost entirely excluded.

Germany's sudden intrusion into Morrocan affairs had probably been entirely unforeseen by the French government, which was devoting its energy and attention to the passage of the bill for the separation of church and state, the most radical act of legislation of the whole period of the third republic, and one which threatened to convulse the nation.

On April 27, the *Matin* published a violent arraignment of Germany's attempt to disturb the peace of Europe, and as this article was thought to have been inspired by M. Delcassé, the Foreign Minister, there was a sudden panic. The situation remained very tense for about six weeks. But it is noteworthy, as indication of the increasing disinclination for war in France, that the policy of M. Delcassé in the foreign office encountered much severe criticism.

Finally, at the suggestion of Germany of course, the Sultan rejected the reform proposals of France, and asked for a conference of the signatories of the Madrid Convention to determine what measures were necessary for improving the administration of Morocco, and how they were to be introduced. This brought the crisis to its decisive point; for Germany promptly accepted the Sultan's proposal, and announced in a note to the French government her view that the questions could be settled only by a conference. It is reported that the German government conveyed through some appropriate diplomatic channel its determination to appeal to arms, if necessary, in support of its position. M. Delcassé's colleagues were unwilling to carry the matter to this extreme point, and forced him to resign, June 6. Three days later, the Kaiser honored Count von Bülow by conferring upon him the title Prince (Fürst), a reward, as it would seem, for the conspicuous diplomatic victory which he had won. The French government accepted the conference with certain preliminary stipulations which were accepted by Germany, and embodied in an agreement signed July 10. These conditions were the sovereignty and independence of the Sultan, the territorial integrity of his empire, commercial equality for all nations, and a recognition of the exceptional position of France with respect to Morocco on account of the long frontier in common.

The *Gaulois* published an interview, July 10, in which M. Delcassé explained his leading views on foreign policy, namely, that France and England should form an alliance, and that such a combination would probably surprise the world by bringing about a reconciliation between England and Russia, and thus expand into a triple alliance.

It has been asserted that the action of Germany in the Moroccan Question in 1905 was due solely to the irritation of wounded pride, since she had not been consulted in the agreement between Great Britain and France concerning the Sultan's dominion. But a broader motive is suggested by a general consideration of the nature of German diplomacy, as embodied in her *Welt-politik*, and this position was formulated by Chancellor von Bülow in a subsequent review of the negotiations in the Reichstag. He affirmed that it was a matter of essential interest to Germany that the unapportioned parts of the earth should not be further reduced, and that the roads should not be closed to the extension of German industry and trade in a land which gave rich promise for the future. The demands of the French envoy were incompatible with the maintenance of the *status quo* and equality for all in Morocco. The actual present extent of German interests in the country was not the essential factor. If Germany allowed her interests to be compromised without a protest on this occasion, the nations would be encouraged to act with the same disregard towards Germany in other cases where the stake might be far greater.

On the other hand, M. Rouvier, the French Prime Minister, who had taken the foreign portfolio upon M. Delcassé's resignation, made the following statement in a report on the Moroccan situation in the Chamber of Deputies:

"Our proposals to the Sultan were not calculated to establish in Morocco a state of affairs similar to that of Tunis. We never assumed the alleged mandate from all Europe. Our envoy, M. Taillandier, performed his function with absolute correctness, which infringed neither the sovereignty of the Sultan, nor the rights of other powers, as based on treaties. We shall confine our policy to the same limits. The question to be submitted to the conference is simple. Every power has rights in Morocco which shall be respected—but it is our duty to indicate to the conference the nature and peculiar importance of our interests."

The Moroccan conference convened, January 16, 1906, at Algeciras, a Spanish port situated on the Strait of Gibraltar, and was composed of the representatives of Morocco, Spain, France, Belgium, Great Britain, Netherlands, Sweden, Russia, Germany, Austria-Hungary, Italy, Portugal, and the United States. The most important questions for deliberation were those regarding the organization of a reliable police system and the establishment of a state bank. In the discussion concerning both of these necessary institutions, there was a serious divergence of views between the French and German representatives; since the former claimed for France a prominent or controlling share in their management. A compromise was finally effected mainly by the moderating influence of Austria-Hungary, which had no direct interest in Morocco. Spanish officers were to be associated with the French officers in the command of the police in the parts where impartiality was especially required on account of international commerce. But the general inspector of the entire police system was to be appointed by Switzerland, supposedly the most disinterested power. Of the capital for the proposed national bank, three-fifteenths were to be furnished by France, and one-fifteenth by each of the other signatory powers. The fact that Italy supported France in this conference, not Germany, was the subject of much comment, and was interpreted in some quarters as an indication of the weakening of the Triple Alliance. Italy had originally been impelled to form the compact with Germany and Austria by resentment at the annexation of Tunis by France, and fear that France would extend her dominion farther along the North African coast. Since that time, the jealousy of Italy and Austria as joint rulers of the Adriatic had increased, and on the other hand France had probably engaged herself secretly to recognize Italy's dominating interest in Tripoli. It must suffice for the present to make note of this ominous phenomenon.

The great importance of a study of Germany's policy in Morocco consists in the fact that it reveals the most prominent feature of German *Welt-politik*. For, at a time when Russia was almost powerless to interfere, and the *Entente Cordiale* had as yet given no certain proof of its firmness, we may fairly assume that Germany's demands were a specific indication of the kind of place which she required for herself "in the sun." Germany demanded no enlargement of her colonial territory, but only the maintenance of the "open door," that is, of impartial treatment for all commercial nations. Great Britain was the only colonial power which had consistently adhered to the practice of treating all alike, and even in the British Empire, the movement for imperial federation, and the establishment of trade preferences seemed to threaten the "open door." A consideration of the movement of German population supports the theory that the principal object of Germany's world-policy was to maintain the "open door." The tide of German emigration reached its highest level in 1881 when 220,902 persons departed from the Fatherland, and began to recede at once, falling to 149,065 in 1884, when the schemes for a German colonial empire were fairly launched. Emigration varied in inverse ratio to the growth of the empire. One might almost be tempted to affirm that the very existence of German colonial territory exerted a discouraging influence on emigration. A comparison with the corresponding movement of British population affords a striking contrast:

	Germany		United Kingdom	
	Population	Emigration	Population	Emigration
1881	45,234,061	220,902	35,019,918	243,002
1911	64,925,993	22,690	45,369,090	440,821

Germany was willing to run the risk of war in support of equal opportunities for her industries and commerce in the parts of the earth which had not been partitioned among the Great Powers. But why should we assume that land hunger has been a controlling motive in her policy, when the colonies which she possesses have been an economic burden, and Germany, as statistics show, has not an excess of population for colonization? During recent years, in fact, the immigration into Germany has more than counterbalanced the emigration from the country. Germany had found an opportunity within her own territory to employ the bulk of her increase in available labor as well as capital. The quotation from the Kaiser's Bremen speech,—"*Aussen hin begrenzt, das innere unbegrenzt,*" which has been freely translated above,—"Confined without, but unrestricted in its capacity for development within," appropriately symbolises the economic position of the German Empire in connection with the guiding motives underlying the race of the nations in securing colonial territory.

An incident happened in the autumn of 1905 which revived for a time the excitement of the situation in the preceding spring, and contributed in part to the increasing current of international animosity. In October, the *Matin* published a "disclosure" by M. Delcassé of the considerations which had prompted his policy as foreign minister. According to this report, he had received Great Britain's promise to support France in the Moroccan affair, even in the event of war, by mobilizing her fleet for hostile action, and by disembarking 100,000 troops in Schleswig in order to seize the Kaiser Wilhelm Canal. This important waterway extends from Kiel to the Elbe below Hamburg, connecting the Baltic and North Seas, and permitting the Germans to employ their naval forces interchangeably in either body of water.

This article had a sensational effect in Germany, and the rather perfunctory way in which the French government denied its truth was not calculated to allay the general feeling of uneasiness. The *Cologne Gazette* observed that it made little practical difference whether the disclosure were true or false. M. Delcassé evidently believed that England had offered aid, and a delusion might have had just as lamentable results as a reality. In fact, M. Delcassé's conviction had all but precipitated Europe into a war, which would have been the most horrible catastrophe that the imagination could conceive. Other journals denounced British perfidy, and the middle class generally believed that Great Britain was awaiting a favorable opportunity to attack them.

The critical events of 1905 were reflected in the renewed

naval exertions of the leading nations. The laying down of the keel-plate of the British battleship *Dreadnought*, October 2, 1905, inaugurated a new era in naval construction. She was of 17,900 tons, and her engines of 23,000 horse-power. Her speed was twenty-one knots. Her armament included ten twelve-inch guns. She was launched February 10, 1906, and made her trial trip October 1, 1906, a year, less a day, from the beginning of her construction, which was considered a record in speed.

The collapse of Russian military power gave Germany an unusual degree of assurance respecting her relative strength on land so that she could divert a larger portion of her resources to increasing her navy. This circumstance, together with the progress of the British navy, and the general political situation—the suspicion, for instance, aroused by Delcassé's disclosure—induced the German government to urge the acceptance by the Reichstag of a measure greatly accelerating naval construction. The new navy bill for revising the program of 1900 increased the number of units, and the size of the larger ships, and advanced the date for the completion of the entire navy, as planned, from 1920 to 1917.

The number of torpedo-boats was to be augmented from ninety-six to one hundred and forty-four. Six cruisers of a special type, nearly as large as battleships, rejected in 1900, were introduced into this bill. There was to be a large increase in the tonnage, and consequently the cost, of the ships that remained to be constructed according to the Naval Bill of 1900. The average cost of a battleship would be raised from $5,827,000 to $8,760,000, and that of a cruiser of the larger class from $4,560,000 to $6,600,000. Two of the ships, whose keels were to be laid at once, were to be of "Dreadnought" size. The plan for future construction as thus amended would involve the addition annually of two

and author of historical works on Morocco.

battleships, and of one large and two small cruisers from 1906 until 1910, of one battleship annually from 1911 until 1916, of eight large and fourteen small cruisers altogether, from 1911 to 1917, and of two battleships in 1917. The total naval expenditure, recurring and non-recurring, for the twelve years covered by the bill was estimated at $890,695,200.

Herr Bebel immediately protested, saying: "I am unable to see what other object this agitation could have than to arm for a war against England. The German navy is not required to fight France and Russia, as the latter power will be paralyzed for many years to come by the effects of its war with Japan, while a war with the United States is out of the question." The figure of the veteran leader of the Socialists raising his warning voice in the Reichstag every time increased naval construction was demanded to predict the inevitable consequence of the government's course recalls the dramatic dignity of an old Hebrew prophet.

The naval bill was passed, May 20, against the opposition of the Social Democrats and some of the Radicals. It is significant that extremists in Germany were not satisfied with this bill as brought in by the government. Their chauvinistic views were most effectively represented by the Pan-German League and the Navy League. The preposition "pan" in such titles as Pan-German should convey the aspiration for the union of all the elements of the nationality or ethnic group indicated by the other part of the word. But the real aim of the Pan-German League was undoubtedly more comprehensive. They wished to have Germany acquire a much greater colonial empire. The extremists demanded that the time for completing the navy, as planned, should be advanced to 1912, instead of 1917. This would have been an unmistakable challenge for absolute mastery on the seas. The government emphatically disclaimed any participation in this view formulated by the Navy League. There was a temporary period of discomposure in the counsels of the society, and its president resigned. But it soon recovered its ardor, passed the million mark in membership, and was a powerful auxiliary in obtaining the passage of another naval bill in 1908 providing a further acceleration of activity in construction.

The German Navy Law of 1906 was the signal for the beginning of the critical stage of the race in naval armaments. Germany set the pace, and Great Britain substituted for her earlier "two-power standard" a scale of construction calculated upon the basis of Germany's activity, endeavoring to preserve a margin of superiority sufficient to guarantee her safety and independence. But the annual sums spent in Great Britain on naval construction do not show the effect of the ever increasing German competition until 1909. In fact, during the previous four years the annual appropriations diminished, a proof of the pacific tendency of the Liberal government, which was severely denounced by their political adversaries as a betrayal of the interests of the country. The following comparative table of annual expenditure for naval construction in the two countries will show whence the impulse to greater exertion came:

Years	United Kingdom	Germany
1905	$54,874,000	$22,940,000
1906	52,776,000	25,113,000
1907	44,843,000	28,727,000
1908	42,088,000	34,882,000
1909	54,564,000	49,460,000
1910	64,559,000	55,369,000
1911	73,210,000	59,536,000

Although it is not the design of the present volume to discuss the details of military and naval strength, the frenzied competition in naval armaments is so largely responsible for the feeling of mutual apprehension and suspicion which unhappily took root in the two countries, that a summary analysis of their respective elements of sea-power at significant points in the race is indispensable for an intelligent understanding of the international attitude in the period directly preceding the war. The following table is intended to provide the necessary data for a comparison of the progress of the two fleets in vessels of the more important classes during the crucial years, all ships of the Dreadnought size, or larger, being grouped under the heading "Dreadnoughts":

	United Kingdom			Germany		
Year	Armored Cruisers	Pre-Dreadnought Battleships	Dreadnoughts	Armored Cruisers	Pre-Dreadnought Battleships	Dreadnoughts
1906	28	49		7	19	
1910	35	53	10	9	24	5
1914	34	40	31	9	22	21

The battleships of the Dreadnought type, or larger, of which none were in commission at the beginning of 1906, have assumed a position of such supreme importance that the contest for naval ascendancy has confined itself mainly to this class. Such vessels, variously classified as super-dreadnoughts, dreadnoughts, and battle-cruisers, are called collectively capital ships, or capital naval units. We must not forget, however, in comparing the naval exertions of the two countries on the basis of their annual expenditure for construction that Germany can devote a larger proportion of her appropriation to ships of the largest size, which are chiefly intended for aggressive action, and that she can concentrate a larger part of her fleet in the North Sea, because Great Britain has a very much larger merchant marine and more vital interests to defend in all parts of the world. Therefore, a continuation of equal appropriations for new ships by the two countries would result eventually in naval superiority for Germany.

The diminishing margin of British naval ascendancy resulted in a feeling of panic in England, when it became evident that Germany was only about a year behind the Mistress of the Seas in the construction of the largest types of battleships, and might by a stealthy acceleration of activity overtake her. This situation drove the Liberal government very unwillingly to appropriate much larger sums for naval construction, although this necessarily reduced the scope of their social legislation, to which they were most earnestly devoted.

When the British budget had been presented in 1908, the naval program provided for seven dreadnoughts and three super-dreadnoughts to be in commission in 1910. The German program at that time provided four dreadnoughts and one super-dreadnought for 1910. Then the new German Navy Law of 1908 prescribed the laying down during the year of the keels for three more dreadnoughts and one super-dreadnought. On this basis the British government assumed that the Germans would have nine battleships of the largest types ready in the spring of 1911. Accordingly, parliament authorized two more battleships, so as to have twelve in 1911 with which to confront the German nine. Later it appeared that the commencement of the construction of the next German set of four great battleships had been anticipated somewhat, so that the Germans might possibly have thirteen capital ships by 1911.

Accordingly the British government asked in the budget

Agadir. The building on the crest of the hill is the citadel, those half way up comprise the walled village of Founti.

of 1909 for authority to lay four keels for ships of the largest types, so as to have sixteen of them in 1911 with which to face Germany's thirteen. But at the same time the British government informed the House of Commons that the length of the interval before the next German keels were laid and the period of construction might be reduced, so that the Germans might have another set of four ships in readiness as early as the spring of 1912, which would give them seventeen altogether at that time. Therefore, they asked authority to take the necessary steps for the construction of four more ships of the largest type during the year, if it should appear necessary.

Prince von Bülow repeated in the Reichstag the government's view that German construction of new ships was determined by Germany's own requirements without reference to competition with any individual power, and that there was no intention of accelerating the legally established rate of construction. He affirmed that they would not have the thirteen large ships before the spring of 1912, and Minister of the Marine von Tirpitz made the same declaration. The British government recovered a feeling of somewhat greater assurance, and decided that the keels of the four supplementary ships did not need to be laid until April, 1910, which would assure their completion in March, 1912.

The Germans had perhaps expected a modification of British foreign policy when the Liberals came into power in December, 1905. But in reality, the Liberal government drew even closer to France, and began to cultivate a cordial understanding with Russia. At the Algeciras Conference the Russian and British representatives discussed in a friendly and informal way the Anglo-Russian position, and each party became satisfied as to the honorable intentions of the other. Then in 1907, Great Britain and Russia formulated an agreement, by which all their differences in

Asia were settled. Persia was to be divided into three parts, one to be a Russian sphere of influence, another a British sphere, and a third lying between these two a neutral zone. The Russian government pledged itself not to carry on any negotiations of a political nature with Afghanistan except through the medium of Great Britain. On the other hand, Great Britain bound herself not to interfere in the domestic affairs of Afghanistan. Both parties, moreover, promised to hold aloof from Thibet. It is a commentary upon the character of international relations that an agreement to adjust misunderstandings is immediately interpreted as a combination with an aggressive purpose. The reconciliation of Great Britain and Russia was looked upon with serious apprehension in Germany. Especially the following year, when King Edward VII had a personal interview with Alfonso XIII at Cartagena and one with Victor Emanuel III at Gaeta in the spring, and was splendidly entertained at Reval by Nicholas II, Tsar of Russia, in the summer, German anxiety increased, and the Germans were convinced of the existence of a Machiavellian project to surround them with a ring of possible enemies.

The Russian policy of the Liberal government in England is variously judged. Its friends look upon it as an honorable recognition of the favorable transformation of the Russian government, while hostile critics condemn it as a debasement of liberal principles. From 1908 the existence of the *Triple Entente*, or cordial understanding between Great Britain, France, and Russia, has been a chief factor in European diplomacy. The formidable naval program of Germany had not been without some influence in promoting international amity, although not precisely along the lines intended. The increase of annual appropriations for naval construction in Germany from $11,784,000 in 1898 to $20,778,000 in 1904 had been a potent influence in bringing about the *entente* between Great Britain and France, and the further advance to $34,882,000 in 1908 probably stimulated very much the inclination of British and Russian statesmen to effect a settlement of the differences between the two nations.

The present chapter would not be complete without mentioning an occurrence of the autumn of 1908, which throws an interesting light upon Anglo-German relations, and itself probably influenced in some measure their course. The *Daily Telegraph* of October 28 published an interview "from a source of such unimpeachable authority" that it could "without hesitation commend the obvious message which it conveys to the attention of the public." It was made up of a number of statements made by the Kaiser, who had sanctioned their publication. The aim of the interview was to prove the Kaiser's sentiments of cordiality for the British nation. The Kaiser deprecated the suspicion which he encountered in trying to bring the nations together. The English themselves, he said, made his task of reconciliation difficult, and as he had to contend against an unfriendly sentiment toward England among a large part of the German people, English prejudice was all the more distressing and unjust. He went on to show that his actions proved the sincerity of his many public assertions of friendship. Thus at the time of the South African War public opinion in Germany was very hostile to England. "But what of official Germany?" he asked. "Let my critics ask themselves what brought to a sudden stop, and, indeed, to absolute collapse, the European tour of the Boer delegates who were striving to obtain European intervention? They were fêted in Holland; France gave them a rapturous welcome. They wished to come to Berlin, where the German people would have crowned them with flowers. But when they asked me to receive them, I refused. The agitation immediately died away, and the delegation returned empty-handed. Was that, I ask, the action of a secret enemy?" The Kaiser went on to say that when the struggle was at its height, France and Russia invited Germany to join with them in calling upon Great Britain to put an end to the war, and that he not only refused to associate himself with this plan for humiliating England, but communicated his reply to the English sovereign in a telegram, which is preserved in the archives of Windsor Castle.

The publication of this interview aroused unusual excitement in Germany, and all parties in the Reichstag insisted that guarantees must be secured that the Kaiser would refrain in future from his practice of expressing himself on the most weighty matters without previous deliberation with his ministers. The Chancellor, Prince von Bülow, undertook to present the view of the Reichstag to the Kaiser, and in a speech before that body a few days later, he assured them of his firm conviction that the Kaiser "would henceforth maintain the reserve that is equally indispensable in the interest of a uniform policy and for the authority of the crown." There was a suspicion in some quarters that the account of a proposal for intervention in the Boer War was a deliberate fabrication calculated to undermine the cordiality of Great Britain for her *entente* associates. But it is more in keeping with the Kaiser's temperament, and the unstudied tone of the interview, to suppose that it was all due to an impulsive outburst. If this is accepted as proof of the Kaiser's sincerity in the interview, we must admire the courage and independence with which he defended an unpopular position in the interests of moderation and good judgment.

Like many other international congresses, the Algeciras Conference failed to make a radical, definite settlement either on account of timidity, or by reason of its eagerness to

The German cruiser *Berlin,* which relieved the gunboat *Panther,* off the closed port of Agadir, Morocco. On July 1, 1911, on orders from the German government the *Panther* anchored off Agadir and the Powers were notified that this was done to protect the lives and interests of German inhabitants from an uprising in the Sus, which the French claimed did not exist.

establish an acceptable compromise between the discordant interests. The Act of Algeciras was too limited in scope to provide an enduring basis for dealing with the difficulties in the situation in Morocco. It proclaimed the sovereignty of the Sultan and the independence of the country, and established a native police force under foreign officers. But what if the Sultan should be powerless to assert his sovereign rights and the police incapable of stemming the rising tide of anarchy? In 1907, Mulay Hafid, brother of Abdul Aziz, who was then reigning, was proclaimed Sultan by the warlike tribes in the south. A civil war ensued, and Abdul Aziz was compelled to retire to Rabatt, whence he could more easily flee the country. In the mean time the fanaticism of the Mohammedan population was inflamed and several French and Spanish were killed in Casablanca, July 31, 1907. In consequence of this, French and Spanish marines were landed there, and only after some sharp fighting succeeded in restoring order. The French and German accounts of these events are at variance, the former representing the natives as impelled to these deeds of murder by their own unprovoked ferocity, the latter representing them as aroused to bloodthirsty actions by the aggressive attitude of the French. Soon the French found themselves compelled to occupy a number of places in Morocco.

Mulay Hafid reached Fez, June 7, 1908. He had already sent to Paris, London, and Berlin to obtain recognition by the respective governments. Abdul Aziz was decisively defeated in August and was compelled to take refuge on French territory. The German government advocated the recognition of Mulay Hafid, and sent their consul, Dr. Vassel, to Fez, although the signatory powers, including Germany, had withdrawn their representatives a short time before to avoid the appearance of recognizing the pretender. The step taken by Germany was apparently intended to secure special favor with Mulay Hafid. He finally received the recognition of the powers upon condition of signing the Algeciras Act. Morocco seemed a little more tranquil, and the French expeditionary force at Casablanca was reduced from 15,000 in March to 9,000 in December.

Rather unexpectedly, as it would seem, the German government concluded an agreement with the French ambassador in Berlin, February 9, 1909, in which Germany recognized the special political interests of France in Morocco, renouncing thereby her policy of upholding the independence and integrity of the Sultan's rule. On the other hand, France pledged herself to maintain the "open door" and to encourage the coöperation of French and German business interests. Perhaps the German government felt that the favor of the new Sultan, which they enjoyed, would counterbalance any unfavorable influence on their commercial interests due to the preponderant position of France. The foreign offices of Europe felt a grateful sense of relief. They fondly believed that the Moroccan question had been settled; but a rude disillusionment awaited them.

In April, 1911, France found it necessary to send a military force to Fez to preserve order. At first, the German government made no protest. Later, they concluded that this was equivalent to abrogating the Act of Algeciras and that the impending "Tunisification" of Morocco would probably begin. Accordingly, in June, the German ambassador in Paris undertook to sound the French government relative to possible compensations for Germany on account of the more commanding position which France had assumed in Morocco. The German government was disappointed that France made no definite offer of compensations; but they were preparing a demonstration which did not yield precedence to the Kaiser's Tangier visit in sensational and dangerous effects. For on July 2, 1911, it was officially announced that the German gun-boat *Panther* had been sent to Agadir, a closed port on the western coast of Morocco. The *Panther* was replaced after a few days by the *Berlin*, a larger vessel. A river of some consequence has its outlet at this point. The *hinterland* of Agadir is fertile in parts and contains mineral deposits. Agadir Bay is said to be not unsuitable for a naval station. Some semi-inspired sheets declared that Germany wanted a harbor in Morocco; but the German government maintained the transparent fiction that the presence of a war-vessel was required in the harbor of Agadir for the protection of German subjects, although some of the opponents of the government's policy declared that there were no Germans there to protect. Evidently, Germany was determined not to be thrust aside without compensation, if Morocco was fated to pass under a French protectorate. Therefore, on July 15, Germany proposed that France cede to her the French Congo from the sea to the Sanga River, and renounce in her favor the French contingent claims to the former Congo Free State, in case Belgium should ever vacate it. This proposal was not accepted by France.

One motive which may have influenced Germany in forcing an issue in Morocco was the desire to test the strength and scope of the Anglo-French *entente*. The British government promptly took an occasion to declare their attitude. For this is the significance generally ascribed to a speech delivered by Mr. Lloyd George at a dinner given by the Lord Mayor at the Mansion House, July 21. He said that it was essential to the highest interests of the world that Great Britain should maintain her place and her prestige among the Great Powers, since her influence had often rendered invaluable services to the cause of human liberty. He was disposed to make great sacrifices to maintain peace, but if a situation were to be forced upon Great Britain in which peace could only be secured by surrendering her great and beneficent position won by centuries of heroism and achievement, by allowing her to be treated, when her vital interests were at stake, as if she were of no account in the family of nations, peace at that price would be an intolerable humiliation.

The expression of these sentiments was doubtless intended to convey the declaration—and in this sense they were certainly interpreted—that Great Britain was prepared to stand by France. It therefore had the effect of a threat in Germany. For nearly two months the nations were on the verge of war. But at last Germany accepted a compromise. According to some, the decisive factor was Germany's financial inability in September, 1911, to mobilize for a warlike demonstration.

Two treaties were signed by the German and French governments, November 4, 1911. The first countenanced the virtual establishment of a French protectorate in Morocco, with the stipulation for equal rights for all nations in trade, customs duties, and mining and railway concessions. The second provided for the cession by France to Germany of a territority in the French Congo, containing 107,270 square miles, to form an eastward extension of the German province of Kamerun, and a cession by Germany to France of 6,450 square miles of the upper part of the same colony of Kamerun.

This conclusion of the troublesome Moroccan Question was very severely criticized by a large part of the German people. The discussion of it in the Reichstag was made the occasion for an attack on the government's policy by all parties. Although Chancellor Bethmann-Hollweg declared in the Reichstag that the attitude of the German government had been consistent throughout the Moroccan controversy, and not modified by any circumstance, there was a general impression that they had receded in the face of the British government's scarcely veiled threat, and the resentment against England was very strong. The extremely bitter attitude found expression in a speech of Herr von Heydebrandt, the Conservative leader, during the discussion of the government's policy in the Reichstag, November 9. The following selection may serve for illustration:

"The Imperial Chancellor has declared that foreign governments were informed in advance that we expected no territorial concession (in Morocco). If, after the British government had been informed of this attitude, a speech, such as Mr. Lloyd George's, was delivered as a deliberate expression of the position of the entire cabinet, it was a threat, a challenge, an insulting challenge. Such speeches are not so readily excused as dinner speeches. They are a most singular kind of dinner speech, which the German people will not put up with. I can well believe that the English are willing to let these matters pass into oblivion, since they failed to involve France and Germany in a war which would have brought them no disadvantage. But we Germans have not forgotten them; and we ask ourselves whether we have been dreaming, or these things have actually taken place. Now we know where our enemy is; like a lightning flash in the night these events have revealed to the German people where its real foe is lurking." The Crown Prince, who had left his military duties at Dantzig to attend this session of the Reichstag, created a sensation by applauding the more emphatic denunciations of the government's policy. On the following day the Chancellor conveyed a rebuke to the extremists of the Heydebrandt stamp, saying that "such words with regard to a foreign state with which we stand in normal relations may serve party interests, but do injury to the German Empire."

We have considered in these two preceding chapters some of the most important indications of Germany's relations with her neighbors, from the resignation of Prince Bismarck to the termination of the Moroccan controversy in 1911. We may suitably pause here for a few moments to summarize the general conclusions to which this evidence leads. But in the first place let us recall the conspicuous features of the German political organization, that the Prussian method of voting is the corner-stone of a system of political inequality and class exclusiveness, but yet, paradoxical as it may seem, that Germany offers models to the world of efficient administration and beneficent socialistic legislation. Germany is almost the only country in the world which has made no substantial change in its constitution since 1871. All the others have modified their fundamental laws in a liberal direction. The relative stability of Germany's institutions is a feature to be taken into account. Her government aims to hold itself aloof from parties. It does not identify its policy with the views of extremists.

The international position of Germany down to 1890 was very strong; she enjoyed almost a military hegemony in Europe. It is true that the black shadow of the French desire for revenge was never entirely absent from the horizon; but on the other hand Germany had formed the Triple Alliance, and France was isolated. Germany cultivated friendly relations with Austria and Russia at the same time. Yet the inevitable antagonism between Austria and Russia was a disturbing factor in Russo-German friendship, and it is doubtful whether even Bismarck could have preserved this threefold understanding intact so as to prevent the establishment of the Dual Alliance of Russia and France. There was still, however, a balance of forces between the Triple and Dual Alliances, and the position of Germany remained tolerably secure. This security would be endangered if the forces of the Dual Alliance received an accession in strength, unless the forces of the Triple Alliance were increased in like measure. Accordingly, the position of any great power standing outside the two alliances would inevitably become very important, and hence Anglo-German relations claim our chief attention. The relations of Great Britain and Germany were cordial in 1890, when Great Britain ceded the island of Heligoland to Germany, and the latter in return gave up her protectorate over Zanzibar. Unfortunately, this state of friendliness was transformed into one of suspicion and distrust a few years later. The responsibility for this deplorable alteration has been charged in part to the Kaiser, but probably with injustice. It is true that some of his speeches have made an unfortunate impression abroad, creating a feeling of uneasiness. The words of monarchs and statesmen pass into circulation with a large premium of implied significance. But in spite of some apparent evidence to the contrary, the Kaiser has probably been a sincere friend of peace, especially with England. The tension in Anglo-German relations was one of the results involved in a fundamental change in the general attitude of Germany.

Soon after the accession of William II, Germany enlarged definitely the range of her foreign policy. She adopted the so-called *Welt-politik* to which she has adhered until the outbreak of the war. The last two chapters have been devoted chiefly to an illustration of the nature of this new attitude, and the evidence seems more than sufficient to show that this "world-policy" embraced two cardinal principles, the maintenance of the "open-door" wherever it existed, and no reduction of the unallotted parts of the earth without Germany's consent as one of the great commercial powers, or without reasonable compensation, that is, corresponding territorial advantages for Germany. This was virtually the expansion of the principle of European equilibrium into that of a world equilibrium, which was surely reasonable, because the progress in means of communication had brought all lands so much nearer. Germany was only formulating for herself a diplomatic doctrine which was justified by her own economic situation and the general economic conditions in the world at large.

Thus the fundamental element of German foreign policy was not unfair. The suspicion of the other powers was chiefly due to the manner in which this policy was asserted, which produced very unfortunate results. In the first place, the Germans enlarged their fleet with such rapidity that Great Britain quite naturally believed that her position as a great power was threatened, and for this reason associated herself with France and Russia in the *Triple Entente*. Then, the foreign intercourse of Germany was frequently conducted in such a tactless, peremptory fashion that she forfeited the sympathy of even neutral nations while supporting demands which in themselves were not unreasonable. In short, Germany's general policy was one of fairness, but her practical methods were unfortunate, and, above all, the tremendous expansion of her naval power was the most threatening potential cause of war that we have thus far encountered.

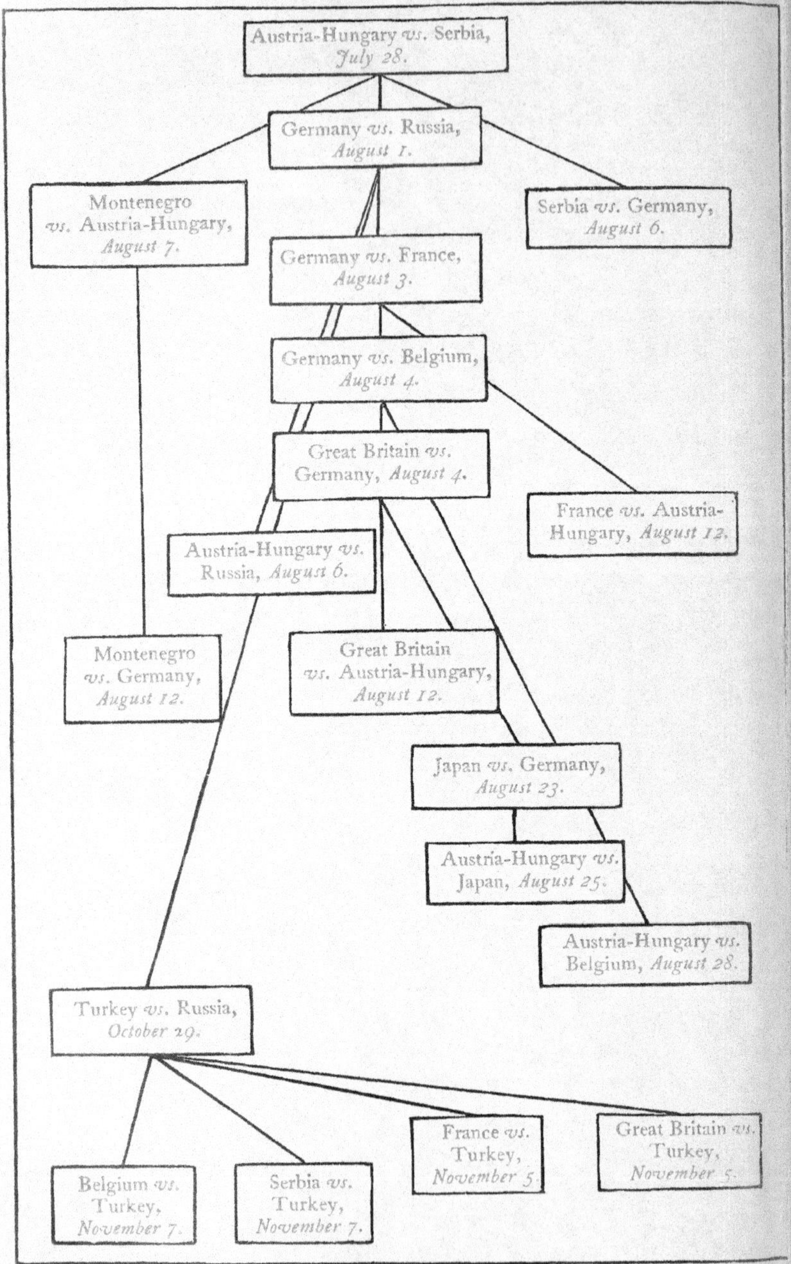

Plan showing dates and causal relationship of the various acts by which a state of hostilities was inaugurated between the different belligerents. *Italy's declaration of war against Austria-Hungary and Turkey, that of Bulgaria against Serbia, and those of Great Britain, France, Russia, and Italy against Bulgaria, which occurred after this chapter had gone to press, and which are not so definitely related to the earlier initiatory acts brought together above, will be treated later. Confusion of dates exists as to some of the "declarations," as the day of their dispatch is not always that of presentation.*

CHAPTER V

The Unavoidable Collision in the Balkans

The series of declarations of hostilities in the world-war. All indications point back to the Balkan Conflict of Slav and Teuton. Geographic and historic conditions in the Balkans. The Southern Slavs. Russia's interests in the Balkans; their origin. The traditional hostility of Russian and Turk. Pan-Slavism. The Russo-Turkish War of 1877–8. Treaty of San Stefano. Treaty of Berlin; its inherent injustice and folly. Russia's vital interest in the Bosphorus and Dardanelles water-way. Germanic penetration in the Balkan peninsula; Austria's motives and intentions; Germany as Turkey's "disinterested friend." Anatolian Railway. The Kaiser's memorable visit in the Turkish Empire in 1898; his protection offered to 300,000,000 Mohammedans. The Deutsche Bank and the Bagdad Railway. Herr von Gwinner. British preoccupations. Macedonian question. Young Turkish party. Revolution of 1908 and Turkish constitution. Annexation of Bosnia and Herzegovina by Austria-Hungary. Serbia's resentment and submission. German influence and the Young Turks; difficulties for German diplomacy. Significance of Italy's campaign in Tripoli. Balkan league; Balkan War and policy of the Dual Monarchy. Second Balkan War, and Treaty of Bucharest, August 10, 1913. Austrian chagrin at the outcome; the effect upon Teutonic military calculations; increase in armaments in Germany, and consequently in France; three years' obligatory military service in France. Russia and the Balkan situation. The field of Greater Serbian propaganda. Political crisis in Croatia. Archduke Francis Ferdinand; the tragedy at Sarajevo, June 28, 1914; the ultimatum, July 23, 1914.

In our consideration of events and conditions preceding the war, we have thus far noted only potential, or general, causes. Now that we are about to approach the positive, or direct, causes, we may for the sake of definiteness proceed at once to the successive declarations of hostilities, and follow back to their source the continuous lines of causation of which these are the landmarks and the culmination. For the world-war, in a formal sense, is the consequence of a series of inaugurations of hostilities announced in a variety of ways.

It will be convenient to enumerate these different initiatory acts to show how one is dependent on another. We begin with the latest.

Turkey commenced hostilities against Russia on October 29, 1914, and this involved her in war with Great Britain, France, Serbia, and Belgium, the allies of Russia in the Great War.

Austria-Hungary declared war against Belgium, August 28th, and with Japan, August 25th, because each of these powers was at war with Germany, her ally.

Japan declared war against Germany on August 23d in pursuance of her ultimatum of August 16th, and in conformity with her alliance with Great Britain.

Great Britain and France declared war on Austria-Hungary, August 12th, because the latter was assisting Germany, with whom they were at war.

Montenegro declared war the same day against Germany, because the latter was an ally of Montenegro's enemy, Austria-Hungary.

Montenegro declared war on Austria-Hungary, August 7th, in consequence of her alliance with Serbia, who was at war with that power.

Serbia declared war against Germany on August 6th, because the latter, in alliance with her enemy, Austria-Hungary, was engaged in war with Serbia's ally, Russia.

Austria-Hungary declared war on the same day against Russia, because the latter was at war with her ally, Germany.

Great Britain declared war against Germany, August 4th, because the latter had violated the neutrality of Belgium.

Germany commenced hostilities against Belgium, August 4th, because the latter refused to allow the German armies to cross her territory to invade France.

Germany announced that she considered herself at war

with France, August 3, really because the latter was an ally of Russia.

Germany declared war against Russia, August 1, in consequence of the latter's menacing attitude on account of the Austro-Serbian conflict.

Austria-Hungary declared war against Serbia, July 28, asserting that the latter had not given a satisfactory answer to her ultimatum.

As each inauguration of hostilities, subsequently to the first, is related individually to a preceding one, the relationship of all of them can be graphically represented in a plan with connecting lines to show the association of cause and effect.

Wherever we encounter a trail of these events, the landmarks guide us unerringly back to a conflict which arose in the Balkans, a conflict which is commonly regarded to be a struggle between the Teuton and Slav powers for supremacy in that peninsula. The lines in our plan are like the intricate passages in a mine, where the routes, though complicated, lead back eventually to the original shaft, by which one ascends to the daylight. The conflict in the Balkans is the clear initial stage, which serves us as definite starting point for threading the baffling maze in which the successive series of causation is involved.

The antagonism which finds expression in the conflict between Austria-Hungary and Serbia is the central positive cause of the world-war. Of all the other breaches of peaceful relations, it has at least been asserted that each was due to some kind of obligation for rendering assistance, made in anticipation of a war. Only this original declaration of hostilities by Austria against Serbia is universally admitted to have had its own independent cause.

There were serious potential causes for war in Europe, as we have seen, boundaries that did not conform to natural geographic conditions, or which violated national unity, bitterness due to former wars, conflicting colonial aspirations, naval competition, international suspicion. Europe was like a group of warehouses stored with inflammable material, threatening a general conflagration. One of these warehouses suddenly burst into flames, and the fire very quickly spread to the others. But nobody will ever know whether combustion would have taken place in one of the others spontaneously. It is our purpose at present to investigate the actual source of the conflagration.

We must consider the respective claims and interests of the Slav and Teuton races in the Balkan peninsula, where they have become involved in this struggle for supremacy; and we shall direct our attention first to the Slavs, whose interests forestalled those of the Teutonic nations by many centuries, and are based upon the actual possession of the greater part of the peninsula. Some prominent historic and geographic facts are of great importance in explaining the position of the Slavs in the Balkans.

During nearly sixteen hundred years Constantinople has been the capital of empires, and during nearly two-thirds of this long period the capital of empires which were declining in power. This unfortunate condition has been ascribed to the hopeless moral decadence of the nations which successively made ancient Byzantium, on the Bosphorus, the seat of their government. But it would have been an unparalleled injustice of destiny to have arbitrarily inflicted so fair a region with a thousand years of decrepitude, if no natural, or physical, circumstances were partly responsible for it; and, in fact, a careful consideration of the situation suggests that certain geographic factors were unfavorable to the maintenance of a vigorous administration of the adjacent parts of Europe and Asia from the city on the Golden Horn as a center. Not only do the Balkan peninsula and Asia

The Yildiz-Kiosk, the palace of the deposed Sultan Abdul-Hamid is the building on the hi
on the right is the Hamidieh Mosque.

Abdul-Hamid II, Sultan of Turkey. Deposed, April 27, 1909.

Minor not form together a natural geographic unit, particularly with the imperfect means of communication which have existed there heretofore, but even the Balkan peninsula alone does not constitute such a unit. True, it has a maritime boundary on all sides but the north, and there the Carpathians, turning westwards before they reach the Black Sea, define clearly the northern limit of the lower Danubian plain, as far west as the famous Iron Gate. But the opening between these mountains and the sea leaves the plain more intimately linked with Russia than with the remainder of the peninsula. Then, farther westward, Serbia and Bosnia are more accessible from the north, from regions outside the peninsula, than from the other Balkan lands. The most conspicuous natural feature of the peninsula is the Balkan range forming a natural barrier running lengthwise for about four-fifths of the distance from the Black Sea to the Adriatic. In the extreme west the coast ranges terminate the Balkan range by running athwart it. There is no natural geographic center for the peninsula. Its contour favors the development of small nationalities. Italy, whose leading geographic features are not ideally arranged for national unity, did not become united until 1870, although her population had really constituted a single nationality for centuries. In the Balkans, therefore, we are not surprised to find several distinct, though mostly related, nationalities; for the boundaries and contour of the peninsula are less favorable for political unity than those of Italy. It was clearly futile to expect that the Eastern Roman Empire could maintain its frontier at the Danube. The Eastern Romans practically abandoned the Danube in the sixth century, and within a short time the whole interior of the peninsula was overrun by Slavs. The Slavs played a far more important rôle in the history of the Balkan peninsula than the German barbarians in the Italian. Each race failed to secure possession of the capital of the peninsula, and this was in each case one of the causes for the lack of political centralization. The Slavs in the Balkans belong to the southern branch of the race, as distinguished from the Eastern Slavs, the Russians and Ruthenians, and the Western Slavs, the Poles, Bohemians and Moravians, and Slovaks. The principal Slavic nationalities in the Balkan peninsula are the Bulgarians in the eastern part, numbering about 5,000,000, and the Serbo-Croats, including the Bosnians and Montenegrins, in the western part, about 9,000,000. And though the Roumanians and Greeks both claim a more distinguished and ancient lineage, they have received an extensive Slavic element into their national composition. Just beyond the limits of the peninsula, or on its very borders, dwell other divisions of the Southern Slavs, the Dalmatians, about 500,000, and the Slovenes, about 1,500,000 in number.

The statement that the Balkans are not naturally a political unit is true with the conditions of communication which still prevail. But with a more developed system of transportation the physical obstacles to unification may in future be vanquished; for the railway defies the barriers of nature. In any case, the peninsula offers great commercial advantages by reason of its three seas, whilst important trade-routes traverse it, notably the one formed by the Danube, and the land-route from Europe to Asia Minor and the Orient, broken only by the Bosphorus, and rendered famous for all time by the march of the Crusaders.

The Slavic nationalities of the Balkans were submerged beneath the Turkish flood of invasion, but did not lose their identity. They began to struggle for their independence when the Turkish power lapsed into decline; and their big brother, and co-religionist, the Russian, came to their aid.

A combination of ideal and material causes has impelled Russia to push southwards, to assume an aggressive attitude towards the Ottoman Empire, and to intervene in the Balkan peninsula, by virtue of her position as the dominant Slavic power, for the purpose of liberating the Christian peoples of the same race from the tyranny of the Sultan.

Two momentous marriages symbolized the association of Russian aspirations with Constantinople. The marriage of the Greek princess Anna with the Russian grand-prince Vladimir at Kieff in the tenth century signalized the introduction of Christianity, which the Russians, in common with nearly all their Slavic brethren of the Balkans, received from Constantinople. The princess Sophia, niece of the last Greek emperor of Constantinople, brought to her bridegroom, Ivan III, grand-prince of Moscow, in the fifteenth century, the double-headed eagle as imperial emblem, the traditions of the Byzantine court, and the notion that the Muscovite rulers were heirs in a way to the dignity of the Greek emperors, which had been handed down from ancient Rome. As indication of these pretensions the Muscovite princes assumed the proud title Tsar, which is supposed to signify Caesar.

To convince oneself of the inveterate character of Russo-Turkish hostility, it is only necessary to glance down the following list of the periods of warfare between the two nations since the time when Peter the Great gave definiteness to the instinctive aims of Russian policy:

Years.

1683—1700: Russia received the port of Azoff.

1710—1711: Peter the Great was forced to resign Azoff.

1736—1739: France supported Turkey, first recognizing that the curbing of Russia was a European interest. The treaty of peace closed the Black Sea to Russian navigation.

1768—1774: Russia obtained freedom of navigation on the Black Sea, and acquired fortresses on the Sea of Azoff.

1787—1792: Russia acquired the territory between the Rivers Bug and Dniester, including the site of Odessa.

1806—1807: The definite treaty signed at Bucharest, in 1812, established the frontier at the Pruth and lower Danube.

1828—1829: The treaty signed at Adrianople established the independence of Greece, and the autonomy of the Danubian principalities, now Roumania.

1853—1856: The war with Turkey involved Russia in the Crimean War with Great Britain, France, and Piedmont. The Russian frontier was pushed back from the lower Danube, a part of Bessarabia being transferred to Roumania. The navigation of the Danube was made free to all nations and intrusted to an international commission. The Black Sea was closed to Russian war-ships, but this restriction was removed, in 1871, with the approval of Prussia, in return for Russia's friendly neutrality during the war against France.

1877—1878: The severest blow to the integrity of her territory was inflicted upon Turkey in Europe. Russia recovered the part of Bessarabia of which she had been deprived in 1856. Roumania, Serbia, and Montenegro were made entirely independent, and Bulgaria and Eastern Roumelia received autonomy.

1914— ?.

The unsophisticated soul of the Russian people is deeply stirred by primitive and simple impulses, such as religious devotion or fanaticism, and an emotional enthusiasm for racial brotherhood. The only popular wars in Russia are those whose object is believed to be the liberation of the fellow Slavic peoples from oppression. The Pan-Slavic societies utilize these sentimental forces to promote their designs for a closer association of the different parts of the race, in which Russia would presumably be the predominant partner.

Nicholas Petrovitch, King of Montenegro, and Queen Milena at Cetinje in 1910.

Ferdinand I, Tsar of Bulgaria, on the ruins of the fortress of Kavala.

Russian religious societies animated the hopes of the Christian populations which were still under direct Turkish rule, by contributing money for the erection of churches and sending books for schools. An insurrection breaking out July 29, 1875, in Herzegovina, was the first event in an unbroken series, linked by the successive relation of cause and effect, which reaches down to the great world-war of to-day. The insurgents solemnly announced their determination to fight for liberty and union with Serbia, and to die to the last man rather than to submit to the unspeakable yoke which they had endured. The Dual Monarchy was already alarmed at the idea of a Greater Serbia, and suggested mediation by Germany, Russia, and herself for recommending the Christian demands to the consideration of the Turkish government, so as to prevent a perilous subversion of the existing arrangements. The Sultan's reply announcing general reforms was in keeping with the usual deceptive practices of the Ottoman Empire in dealing with the powers. The Porte accepted the Andrassy note demanding definite measures of reform, January 31, 1876, but the insurgents, rendered wary by past experience, did not lay down their arms. A slight movement in Bulgaria was a pretense for turning loose the bloodthirsty Moslem irregular soldiery on the defenseless country, and an orgy of butchery ensued, known as the "Bulgarian atrocities." About 15,000 persons were massacred, and seventy-nine villages were burned. At this time, Sultan Abdul-Aziz was murdered; his successor Murad V represented the uncompromising Turkish party. Serbia declared war, June 30, and from their mountain eyries the unconquered Montenegrins hurled defiance at their ancestral enemy, the infidel. The fact is not without significance that there were pro-Turkish demonstrations in the streets of Buda-Pesth; for Hungarian suspicion of the Balkan Slavs is not a recent development. The negotiations of the Porte were characterized by the usual duplicity and procrastination. The Serbians suffered a severe defeat. Murad V was deposed and his brother Abdul-Hamid substituted for him, destined to a long reign with events of far-reaching significance. The Turkish government enacted a parody for the benefit of the powers, which for consummate satiric genius has not been excelled. The powers demanded reforms; the Turks proceeded to outbid them in magnanimity. They solemnly proclaimed a Turkish constitution, and the summoning of a parliament of representatives of all the peoples in the Ottoman Empire, January 20, 1877. Henceforth all questions concerning the integrity of the empire could be considered only by the legally constituted organs. The Turks were encouraged in all their subterfuges by the conviction that they had a steadfast friend in the Disraeli Conservative administration in England.

Russia alone remained resolute in her attitude toward the Turks. A wave of generous enthusiasm for the cause of the oppressed Christians swept over the country, and the people demanded armed intervention, while the government, reluctant to undertake a war, patiently exhausted the last resources of diplomacy. As early as January 15, 1877, in a secret convention, Austria's consent to Russia's eventual military intervention in Turkey had been obtained by the agreement that Austria might "occupy" Bosnia and Herzegovina.

Although Great Britain abandoned the Turkish government for a time to the punishment which it merited, a limit to her concession was clearly announced, while at the same time her traditional attitude was firmly expressed in a note to the Russian government, May 6, 1877, as follows:

"The vast importance of Constantinople, whether in a military, a political, or a commercial point of view, is too well understood to require explanation. It is, therefore, scarcely necessary to point out that her Majesty's government are not prepared to witness with indifference the passing into other hands than those of its present possessor of a capital holding so peculiar and commanding a position."

The Russian declaration of war reached Constantinople April 24, but owing to the deficient means of communication and unfavorable weather conditions the Russian armies did not cross the Danube until June 27. Roumania had been induced to allow the Russian forces to traverse her territory, and, as an almost inevitable consequence, she united her forces with Russia as an ally. Bulgaria was soon overrun and the Shipka Pass through the Balkans occupied. Then came a set-back. Osman Pasha took up a position at Plevna on the right flank of the Russian line of advance, and made one of the most brilliant stands in history from July 20 until December 10, surrendering at length to a force three times his own. This long delay was followed by a winter campaign. The resistance of the Turks was finally crushed, the Russians poured over the Balkans, their lines converged at Adrianople, January 20, 1878, and they were encamped on the Sea of Marmora, within sight of the minarets of Constantinople, on the 31st, when preliminaries of peace were signed. The Treaty of San Stefano, between Russia and Turkey, was concluded March 3. The terms of the treaty reveal the good sense of the Russian government, and a comparison between its provisions and those of the Treaty of Berlin, which was substituted for it, is most instructive. Every alteration made by the powers in the Russo-Turkish arrangements, as embodied in the earlier treaty, has been fraught with dissension, international antagonism, and bloodshed.

The essential feature of the terms of San Stefano was the creation of a big Bulgaria as an autonomous principality with a section of sea-coast on the Aegean as well as on the Black Sea, and embracing most of Macedonia. Russia was to accept, in place of part of the war indemnity a territory of about 12,000 square miles in Armenia, and her boundary was pushed forward to the Danube by the restoration to her of the part of Bessarabia of which she had been deprived in 1856. Roumania received the Dobrudscha in compensation for this cession of territory, and, together with Serbia and Montenegro, became entirely independent, whilst Bosnia and Herzegovina were to receive autonomy. The treaty was a disappointment to the Christian states of the peninsula, with the exception of Bulgaria, and it encountered the determined opposition of some of the Great Powers, particularly Austria and Great Britain, who demanded that it be submitted to revision by a European congress. Austria-Hungary insisted upon her share in the booty, although she had contributed neither men nor money to the liberation of the Balkans. Her intervention was limited to the exaction of toll, as it were, from the power that actually performed the task, by requiring some sort of promise of compensation. Now Austria mobilized, and assumed a threatening attitude. The war had been costly in lives and treasure. Russia was further embarrassed by the Nihilist propaganda, which had become a serious problem. She adopted, necessarily, a course of moderation, and accepted the demand that the regulation of boundaries within the former Ottoman Empire should be made a subject for decision by the powers.

The diplomats of Europe assembled at Berlin, June 13, 1878, and the ill-starred document containing the results of their deliberations was signed on July 13. At the suggestion of the British envoy, Austria-Hungary was intrusted with the "occupation" of Bosnia and Herzegovina, and thus the hope of the Serbs uniting their entire nationality, with which they had taken up arms in 1876, was cruelly frustrated. The Bulgarians were divided into three distinct parts. The Macedonian section returned under Turkish misrule, destined to be a continual source of vexation, the chief cause of the Balkan War in 1912, and of the ensuing violent contest between the Balkan allies. The remainder of the Bulgarians were separated by a line drawn along the summits of the Balkan Mountains. The section to the south of it was made an autonomous province of the Turkish Empire, called Eastern Roumelia, with a governor appointed by the Sultan, while the section to the north, between the mountains and the Danube, became the Principality of Bulgaria, electing its own ruler, but tributary to the Turkish government. This division was manifestly artificial; and in its arrangements as a whole, the congress appears to have been indifferent to justice, popular desires, and good sense. The territorial acquisition of Russia in Asiatic Turkey was reduced. She received a strip containing Kars, Ardahan, and Batoum, smaller in area than the cession to Austria, if we should reckon Bosnia and Herzegovina as already Austrian. Lord Beaconsfield returned to England from this inglorious assembly with the announcement that he brought "peace with honor."

Whatever may have been the expectations of Russia that she could use the greater Bulgaria as an effective medium for the extension of her influence, the prospect of which had aroused the fear of Great Britain, subsequent events have proved beyond a doubt that a stronger Bulgaria would have served as a bulwark for protecting the independence and peace of the Balkans, by discouraging the intrigues of the powers. The Bulgarian principality as actually established at Berlin, with its very much reduced territory, asserted its independence, and exhibited a remarkable progress. Grateful at first for the sacrifices which Russia had made in their behalf, the Bulgarians accepted Russian advice and assistance. The Sobranje, or popular assembly, elected Alexander of Battenberg as their prince, in 1879, and admitted many Russians into important civil and military posts. But their arrogance alienated the Bulgarians, and the more prominent Russian officials were forced to resign in 1883; and from that time, Bulgaria strove to free herself from Russian tutelage, although she forfeited Russian favor by this policy, which is a fact of great importance in recent Balkan history. By a bloodless revolution at Philippopolis, in 1885, the governor was expelled and the people proclaimed their union with Bulgaria. This first contravention of the Treaty of Berlin was an occasion for much resentment and discord. The Serbs, complaining that it destroyed the equilibrium of power, invaded Bulgaria, but suffered an unexpected succession of defeats, and only the diplomatic intervention of Austria secured for them a treaty of peace reëstablishing their position as it had been before the war, March 3, 1886. The Bulgarian prince, Alexander, resigned in September, perhaps on account of the enmity of Russia, and both the Bulgarians and Eastern Roumelians elected Prince Ferdinand of Saxe-Coburg as their ruler, July 7, 1887, which was equivalent to the union of the two principalities, although the powers did not officially recognize his election until 1896. Under the vigorous administration of his able prime minister, Stambuloff, Bulgaria prospered greatly. The finances were placed on a stable footing, the civil service was made efficient, education was fostered, and railways were built. The campaign of 1885 had established the military reputation of the nation. Bulgaria began to be regarded as a Balkan Prussia. The success and progress of the Bulgarians prove incontestably that the apprehension of Russian domination in the Balkans, as likely to be exercised through the instrumentality of the greater Bulgaria of the Treaty of San Stefano in 1878, was groundless.

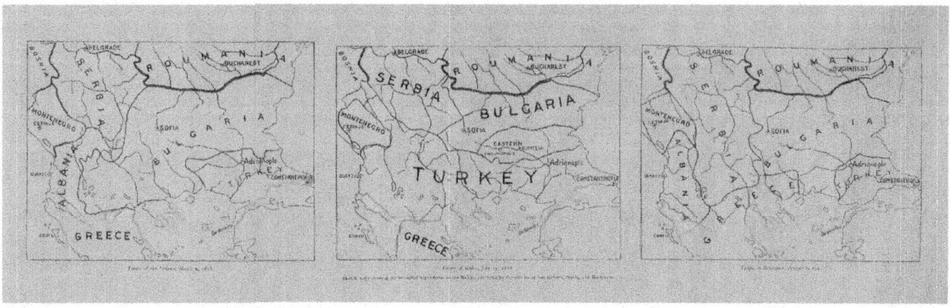
BELGRADE
ROUMANIA
BUCHAREST
SERBIA
BOSNIA
MONTENEGRO
SOFIA
BULGARIA
Adrianople
ALBANIA
TURKEY
CONSTANTINOPLE
GREECE
BELGRADE
ROUMANIA
BUCHAREST
BOSNIA
SERBIA
BULGARIA
MONTENEGRO
SOFIA
EASTERN
Adrianople
TURKEY
CONSTANTINOPLE
GREECE
BELGRADE
ROUMANIA
BUCHAREST
BOSNIA
SERBIA
MONTENEGRO
SOFIA
BULGARIA
Adrianople
ALBANIA
TURKEY
CONSTANTINOPLE
GREECE

But Russia had commercial interests in the Balkan peninsula, and more especially in its adjacent waters, to which her political interests are really subordinate.

The deficiency of the vast Russian Empire in available maritime outlets is a commonplace of geography. The White Sea scarcely counts; and while the Baltic coast has at least one open winter harbor, the center of commercial and industrial gravity, as we have already perceived, is inevitably moving southwards, so that the acuteness of the problem depends in large measure upon the opportunities for external communication in that direction. Russia's most productive natural sources of wealth are chiefly found in the regions which are tributary to the Black Sea. The Black Earth Belt of marvellous fertility, the rich grazing lands of the steppes, the boundless supplies of coal in the Donetz basin, the oil-wells in the Caucasus—the outlet for all these to the great international trade-routes is by the Black Sea and the narrow passage of the Bosphorus and Dardanelles. The area within which this movement of southern Russian foreign trade takes place is funnel-shaped. The products move along converging land or river routes to the Russian ports of the Black Sea, then by maritime lines of communication drawing ever closer, until they reach the entrance of the Bosphorus, where the narrow spout of the funnel begins. The passage-way through the Bosphorus and Dardanelles is essential to the prosperity of the Russian Empire. Let us consider a moment the full significance of this. The population of the Russian Empire amounts to-day to about 175,000,000, and it will be 600,000,000 at the end of the century, if the present rate of increase continues. Russia produces more than one-fifth of the world's wheat, about one-fourth of the potatoes, more than one-fourth of the oats, seven twenty-fourths of the beet-sugar, nearly one-third of the barley, and more than one-half of the rye.

The agricultural development is capable of indefinite expansion. Russia by intensive farming could probably feed the world. All civilized nations are interested in Russian exports. A stoppage of the Dardanelles for a few weeks sets the price of wheat soaring all over the world. It is both unreasonable and undesirable that such a huge productive area as we find in southern Russia should depend upon a precarious right of way, subject to the humor of a possibly hostile power, for access to the great international water-routes.

Constantinople exercises a traditional spell over the imagination of the Russians. They dream of a time when the cross will supplant the crescent on the dome of the mosque of St. Sophia, where once the Greek emperors were crowned. But the practical importance of the city itself for Russia is of rather a negative sort. The charm of an incomparable situation, and the prestige of a mighty historic past maintain Constantinople's superior rank. But it is not the natural distributing point for an extensive territory. The exports from Constantinople are of limited amount. Smyrna in Asia, and Salonica in Europe, have each a more extensive economically tributary hinterland. The enormous bulk of the tonnage of the shipping of Constantinople is deceiving; it consists in large part of steamships of the Black Sea lines which merely touch there in passing. In accordance with a constant economic law, maritime commerce seeks the ports nearest to the sources of production or areas of consumption, which in this instance are in southern Russia. Constantinople and Odessa, the most important of the Russian Black Sea ports, will eventually assume the relative positions of Detroit and Chicago. Constantinople's commercially strategic situation enjoys consideration which is out of all proportion to its own intrinsic, commercial importance. In other words, Con-

Charles I, King of Roumania, 1866–1914, and his wife Elizabeth of Wied known as "Carmen Sylva."

uncle as King of Roumania, October 10, 1914

stantinople, like some important railway junctions, derives a reflected commercial glory from the crossing of two great trade routes, the water-way from the Black to the Aegean Sea, and the land route from central Europe to the interior of Asia, without making any considerable contribution to the importance of these routes. The practical interest of Russia in Constantinople and the Straits is chiefly confined to her concern for the freedom of navigation. If Russia possessed complete freedom for merchant ships and warships alike, and had absolute assurance that liberty of navigation would never be restrained, the actual possession of Constantinople itself would convey very little additional advantage. But, on the other hand, the domination of Constantinople by a strong, and possibly unfriendly power, capable of arresting the commerce of the Russian ports, would be fatal to the prosperity of Russia. The Teutonic combination of powers was establishing precisely such a hegemony in Constantinople, full of peril for Russia. For this reason, we must direct our attention forthwith to the origin and rise of the Germanic influence in the Balkan peninsula.

We have observed that Austria-Hungary by her "occupation" of Bosnia and Herzegovina obtained from the Russo-Turkish War more profit than the victors themselves, provided the domination of alien peoples is regarded as an advantage. By this extension of Austrian authority, the policy called "*Drang nach Osten*," or "pushing eastward," found practical expression. It is difficult to form a definitive judgment regarding Austria-Hungary's advance into Bosnia and Herzegovina. If the government of Francis Joseph was impelled by a desire to find compensation in the Balkans for the loss of its Italian provinces, we must condemn emphatically such an unreasonable determination to seek aggrandizement in utter disregard of the sentiment of the people over whom it insisted upon extending its rule.

It is hardly necessary to state that the people of Bosnia and Herzegovina have never been invited to declare by plebiscite their own inclination respecting their destiny. That method of consulting the popular preference has been employed by the more enlightened Latin nations only, and by them, as it happens, only in instances where the result could be foreseen, and was in conformity with the wishes of the authorities who left to the people in this way the control of their destiny.

Public opinion is always inclined to condemn Austria-Hungary without a hearing. But fairness demands that we consider the Austrian assertion, that these provinces were in a chronic state of disorder, which made them a constant source of annoyance to their neighbor. Moreover, one must candidly admit, that during the thirty years of their "occupation" the Austro-Hungarian government did its uttermost to justify its control, bringing about a wonderful transformation by establishing an impartial and efficient administration, reducing to a coherent system the chaos of laws, and building schools, hospitals, highways, and railways. The Bosnian landscape took on an appearance not unlike that of the parts of Austria where civilization had already existed for many centuries.

The ultimate designs of the Dual Monarchy in the Balkans were probably never very clearly defined, although it has been commonly assumed that the Austrian government hoped in some way to extend their power to Salonica, as an outlet on the Aegean Sea.

While the penetration of Austria-Hungary implied from the first a definite territorial advance, the penetration by Germany was naturally directed towards the attainment of a general diplomatic and commercial hegemony.

With the coming of Mr. Gladstone to the head of the British cabinet in 1880, the Turkish government must have been convinced that it could no longer count on British support. The scathing denunciations of the Bulgarian atrocities by Mr. Gladstone in 1876 were not the transient effusions of campaign oratory. Germany gradually assumed the place of "disinterested friend," which had thus been left vacant. One of the earliest indications of this new element in Balkan affairs was the beginning of the re-organization of the Turkish army under German management. General von der Goltz undertook the task in 1883, and remained in charge of the Turkish military establishment until 1895, and in consequence many German officers were introduced into the Turkish service, whilst many young Turkish officers were sent to Germany for their technical education. Germany began to support Turkish interests diplomatically, and to work for, and believe in, the economic regeneration of the country. The Germans adopted a tactful attitude in their relations with the Turkish government. It was easier for Germany to pursue an apparently disinterested policy, because, unlike Austria, she was geographically remote from Turkey.

Abdul-Hamid's personality combined the fascination of manner, which might naturally be developed amid the intrigues of an imperial harem, with unusual intellectual alertness and tenacity of purpose. He aimed to consolidate his authority by reviving the religious power of the caliphate, which had been combined with the Ottoman Sultanate since 1517, but which had fallen into neglect in recent times. A Pan-Islamic propaganda for stimulating the sentiment of community throughout the Moslem world was a powerful auxiliary in Abdul-Hamid's plan of reinvigorating the caliphate. Abdul-Hamid determined to combat the disintegrating forces, which were undermining the Turkish Empire. His quick perception discovered the cause for the retrogression of Turkish dominion in the easy indifference with which the earlier Sultans had tolerated the existence of the individuality of the subjugated peoples. It was now too late to amalgamate them with the ruling nationality; but Abdul-Hamid, although feigning at times to yield to pressure, usually thwarted the introduction of such administrative reforms as might pave the way for the eventual separation and independence of individual provinces. The revolt and massacres of the Armenians in 1896, when even the streets of Constantinople were drenched with blood, were a consequence of the Pan-Islamic spirit of fanaticism.

Abdul-Hamid restored the old-time despotism. He reduced his ministers to the position of vassals or servants, from which they had risen to practically independent power under his feeble predecessors. He believed that he was peculiarly favored by the stars, and therefore called his favorite palace on the Bosphorus Yildiz Kiosk, or the Star Villa; and there he usually dwelt, in voluptuous oriental seclusion, but retained in his clutches all the cords by which the different parts of the administration were manipulated.

Bismarck was apparently indifferent to the opportunities afforded by the Turkish Empire. His sole aim in the near east was to maintain a useful equilibrium between Austria and Russia. But we have already noticed the unusual combination in the present Kaiser's character of an ideal or romantic vein with a practical appreciation of the forces and factors of the economic world of to-day. He endows with the compelling quality of picturesqueness the sober paths of utility and material profit. Within a year of his accession to the throne, William II paid a state visit to Constantinople, casting a spell, as it were, to facilitate the progress thither of German capital for investment. The intimate association of the German financial system with industry and commerce is a powerful factor in encouraging German enterprise in all parts of the world. In 1888, the

Baron Kolmar von der Goltz, field marshal of the German army (*the first figure in uniform on the left*), reorganizer of the Turkish army.

Deutsche Bank took over a short railway running from Haidar Pasha, near Scutari, opposite Constantinople, out into Asia Minor, and secured a concession for its extension to Angora, and the following year the Deutsche Bank and Württembergische Bank together founded the Anatolian Railway Co., securing a further concession for a line to Konia, four hundred and sixty-seven miles from Haidar Pasha, which was completed in 1896. Herr von Siemens, then chief director of the Deutsche Bank, was the animating spirit in this undertaking. The Deutsche Bank, the twin brother of the German Empire, as it is called, having been born (within the same year) in 1870, is the most prominent of the great German banking institutions which finance enterprises in foreign parts.

In the autumn of 1898 the Kaiser, accompanied by the Kaiserin, made a much more pretentious journey to the Ottoman Empire, ostensibly to dedicate a German Protestant Church in Jerusalem, October 31, 1898, when the Kaiser concluded an address with the following words:

"From Jerusalem came the light, in the splendor of which the German nation has become great and glorious, and what the Germanic peoples have become, they became under the banner of the Cross, the emblem of self-sacrificing Christian charity. As nearly two thousand years ago, so there shall to-day ring out from Jerusalem the cry voicing the ardent hope of all, 'Peace on Earth!'"

A few days later, November 7, the Kaiser declared before a gathering in Damascus that "the Sultan and 300,000,000 Mohammedans, who, scattered over all parts of the earth, venerate him as their Caliph, can ever rely upon the friendship of the German Emperor."

The seemingly contradictory light in which the Kaiser appeared, the unusual character of this entire affair, and the uncertain, but vast, extension of German influence which his words seemed to betoken brought forth much adverse criticism and apprehension. To account in part for the general surprise, we may recall the fact that these sentiments were expressed two years after the Armenian massacres, for which the world regarded Abdul-Hamid as partly responsible, and the year after the victory of the Turks in a war with Greece, whose crown princess was the Kaiser's sister. But above all, to explain the startling effect of his speech in Damascus, we need only regard the territorial position of the 300,000,000 Mohammedans to whom his protection was thus offered, of whom fully 75,000,000 dwell under British, and many more millions under French, authority.

Those who have been admitted to the friendship of the Kaiser are impressed by his open-hearted, sympathetic nature. His amiable enthusiasm shows itself in his public utterances, when he seems to be frequently impelled by his warmth of cordiality, or by the inspiration of the moment, to employ expressions which are liable to an unforeseen interpretation. The Kaiser's intentions on this occasion were doubtless very much less extravagant than many of the extreme constructions which were placed upon them. He desired to draw near to Turkey, and he did not lose sight of the fact that other Mohammedan lands contiguous to the Ottoman Empire offered exceptional opportunities for German trade. But it will be well for us to keep in mind, from this point on, that the Mohammedans inhabit countries which command the most important strategic points in the world with respect to international trade-routes.

The events which we have been relating occurred the same year as the China expedition and occupation of Kiau-Chau by Germany. The Kaiser's *Welt-politik* was evidently moving forward, and events showed that his progress through Turkey, which had been inaugurated by a visit at the court of the Sultan and superb entertainments, was admirably timed for promoting a practical program, and was not barren of results. For within twelve months, a concession was obtained from the Porte for the extension of the German Anatolian Railway to the Persian Gulf. We may note in this connection the fact that the railways in European Turkey were already under the control of the Deutsche Bank. Reflections of the most far-reaching significance are at once suggested. The commercial supremacy of Great Britain, and her empire, were based upon the control of sea-power. The sea-route to India was unquestionably in her hands. But the continental states by the construction of railways might make themselves independent of water-routes. The head of the Persian Gulf was surely not the ultimate terminus of the great route contemplated in the German line. A glance at the map will convince the reader of the expediency, or even inevitableness, of a line of railway communication from the west of Europe to India and China, crossing the Bosphorus by means of car-ferries, following ancient caravan routes, paralleling in effect, and lying between, the Trans-Siberian line of Russia and the all-sea connection of Great Britain. This would offer a line of "peaceful penetration" for German civilization into Asia Minor and Mesopotamia, and even Persia, which might lead eventually to the demarcation of a German "sphere of influence" in these regions. Here was the one suitable outlet by land for the expansion of German enterprise, a long narrow zone, sandwiched in, as it were, between the spheres of Russia and England.

The prospects of railway extension stimulated the Turkish government's interest in the country at the head of the Persian Gulf, which was ruled by practically independent sheichs under the nominal hegemony of the Ottoman Empire. In August, 1901, an attempt was made to land Turkish troops at Koweit, a desirable harbor. But the local sheich appealed to the British Indian government, and a warship was despatched, which thwarted the Turkish design. The news of this incident was, doubtless, the cause of some annoyance in Berlin. It was a palpable declaration that the Persian Gulf was included in a British "sphere of influence."

In 1901, Arthur von Gwinner succeeded von Siemens as leading director of the Deutsche Bank, and became one of the big figures in the economic struggle for a greater Germany. He became the guiding genius of Germany's penetration of the Asiatic territories of the Ottoman Empire. Herr von Gwinner is an admirable specimen of the German captain of finance. He is a scholar, an art connoisseur, a music lover, and above all a prophet and seer. It requires the soaring imagination and vision of a prophet to mobilize and guide to victory the modern forces of finance. Amid the severely prosaic surroundings of his office in the Behrenstrasse, he could invoke the vision of Babylon and Nineveh with their imperial splendor, and the opulent luxury of Bagdad, the commercial metropolis of the earth in the days of Haroun-al-Raschid and the Arabian Nights; he beheld the present desolation of Mesopotamia where nature's lavish resources are neglected by the scanty population; and he could forecast a future when the valleys of the Euphrates and Tigris, brought within five days of Berlin by railway, would be repeopled and cultivated, and dotted with German homesteads.

The definite concession granted by the Porte, March 5, 1903, was the charter of the Bagdad Railway Co., the *Chemin de fer impérial ottoman de Bagdad*, as it was officially called. The capital of the company amounted to 15,000,000 francs, of which only one-half was paid up. But with this insignificant sum as a basis they undertook to construct a line 1025 miles in length from Konia to Bagdad, or 1400 miles altogether if it were extended to the Persian Gulf.

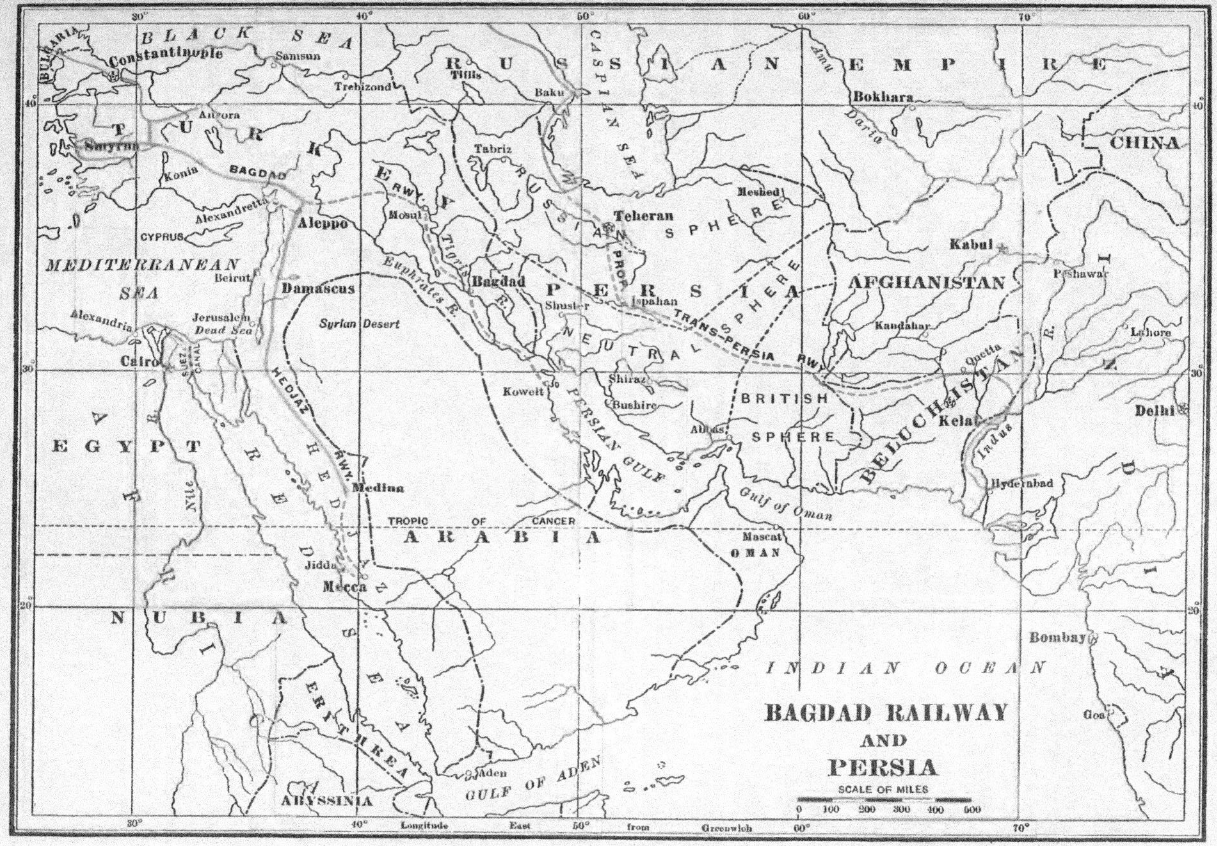
BAGDAD RAILWAY
AND
PERSIA
SCALE OF MILES
0 100 200 300 400 500
Longitude East from Greenwich
BLACK SEA
Constantinople
Samsun
Trebizond
RUSSIAN EMPIRE
Tiflis
Baku
CASPIAN SEA
Bokhara
Amu Daria
CHINA
TURKEY
Angora
Smyrna
Konia
BAGDAD
Alexandretta
Aleppo
CYPRUS
MEDITERRANEAN SEA
Beirut
Damascus
Jerusalem
Dead Sea
Alexandria
Cairo
SUEZ CANAL
Syrian Desert
Mosul
Tigris
Euphrates R.
Bagdad
Tabriz
RUSSIAN SPHERE
Teheran
Meshed
PROP
PERSIA
Shuster
Ispahan
NEUTRAL SPHERE
TRANS-PERSIA RWY.
AFGHANISTAN
Kabul
Kandahar
Peshawar
Lahore
Quetta
BELUCHISTAN
Kelat
Indus
Hyderabad
Delhi
INDIA
Koweit
Shiraz
Bushire
PERSIAN GULF
Abbas
BRITISH SPHERE
Gulf of Oman
Mascat
OMAN
EGYPT
AFRICA
Nile
RED SEA
HEDJAZ RWY.
HEDJAZ
Medina
TROPIC OF CANCER
ARABIA
Jidda
Mecca
NUBIA
ERYTHREA
ABYSSINIA
Aden
GULF OF ADEN
INDIAN OCEAN
Bombay
Goa

The terms of the contract might excite the envy of our most fortunate promoters of public utilities. They show why the public contracts are a subject of such earnest consideration in the semi-civilized countries whose status was discussed from time to time in European diplomatic conferences. The Turkish government agreed to contribute to the railway company an annuity of 11,000 francs per kilometer of line ($3396.80 per mile) for ninety-nine years, and guaranteed minimum operating receipts of 4500 francs per kilometer ($1389.60 per mile). Then, to facilitate the financing of the project, the government turned over its own bonds to the company, as the work proceeded, representing a nominal value of 269,110.65 francs for every kilometer ($83,101.35 per mile). The railway company sold these bonds, for which the government's annuity provided the interest at 4% and redemption at the end of ninety-nine years. Furthermore the company received the privilege of exploiting mines within a distance of twenty kilometers (12½ miles) on either side of the line, and of cutting timber in the neighboring forests, and other valuable rights. At the commencement of the war about $60,000,000 had been invested in the railway, and its completion to Bagdad was expected in 1917.

Herr von Gwinner, as head of both the Anatolian and Bagdad Railway Companies became railway king of the Ottoman Empire. The line of the Bagdad Railway, as surveyed, and largely constructed, starts from Konia, passes near Aleppo, the metropolis of Syria, and through Mosul on the site of ancient Nineveh, and then, descending the Tigris valley, reaches its termination in a splendid new railway station in Bagdad close by the quay of Nebuchadnezzar. The contract for the construction of the line was immediately granted to a German company, and the first section opened for operations in 1904.

The supreme importance of the Bagdad Railway will be realized through its eventual extensions, connections, and branches. In the first place, it was very clearly designed to be the central section of a short route to India and the extreme Orient. For this reason the project excited from the first the concern, and even apprehension of Great Britain. It will be observed in this connection that Russia's attention was absorbed at the time with preparations for her struggle with Japan, so that she had little time to devote to events and developments within the Turkish Empire. The connection with India for the Bagdad Railway was to be effected primarily by an extension of the line from Bagdad to some suitable port at the head of the Persian Gulf, a distance of three hundred and seventy-five miles. The attitude of Great Britain was exemplified by the Koweit affair. The railway project from Bagdad to the Persian Gulf remained unsettled for a long time on account of the suspicion of Great Britain. It was finally agreed in 1914 that this section of railway should be under exclusive Turkish control.

The German penetration of Mesopotamia introduced a new factor into the preoccupations of British statesmanship regarding the security of India. This was very obviously one of the causes of the agreement with Russia in 1907, to which reference has already been made.

The Bagdad Railway added the crowning element of usefulness to a favorite project of the Sultan. Abdul-Hamid planned a railway line from Aleppo and Damascus to the Moslem holy cities, Medina and Mecca, to be constructed by means of the contributions of Mohammedans in all lands, as a practical, conspicuous monument of his régime as Sultan and Caliph, and an impressive symbol of Pan-Islamism. The Hedjaz Railway, as it is called, has been completed from Aleppo to Medina, a distance of about eight hundred miles, and was eventually linked to the Bagdad Railway by a branch of the latter extending to the town of Aleppo.

Later, the Germans obtained a concession for a branch line from Aleppo to Alexandretta, a port on the Mediterranean, where a naval base might some time be established, only thirty-six hours steaming from the Suez Canal.

The military advantages of the Bagdad Railway as the main trunk of an extensive system must have been as patent to the cunning mind of Abdul-Hamid as to its promoters. Its possibilities were so stupendous as to bewilder analysis. It was not only an interior system to facilitate the rapid disposition of troops wherever the coasts might be threatened from the sea; it not only constituted the most important part of a short route to India; but it provided continuous railway connections from Germany to a point within striking distance of Egypt, if only the barriers were lowered in the Balkan peninsula. In other words, it could bring the German military machine almost down to the Suez Canal; and this was, doubtless, the most insidious danger which it held in store for the British Empire. And as though to clench the strategic advantage, the Hedjaz Railway ran far down the Red Sea, along the flank of the great British trade-route, where a site for some future naval base might conceivably be discovered, as a useful ambuscade. Not unconnected with this general program, was Abdul-Hamid's attempt in 1906 to occupy the Sinaitic peninsula, and push the Turkish frontier nearly up to the Suez Canal, a scheme which was thwarted by British initiative in Egypt.

The British Foreign Office was prompt to grasp the dangerous possibilities of this extensive railway development under German control. Its effect on British Indian policy was revolutionary. The English abandoned the traditional idea of keeping India isolated by land, and became convinced that railway communications with their greatest dependency were required. Thirty years ago the cry of alarm was raised in Great Britain, because the Russians were "at the gates of Herat," with the design, as it was believed, of invading India. And now the presence of the Germans at Bagdad, and more especially the possibility that the eastern Mediterranean or Suez Canal might be closed, were causing similar anxiety. To provide a means for the speedy transportation of forces overland to India, Great Britain was constrained to put confidence in the loyalty of her *entente* associate, and an agreement was practically consummated for the construction of a railway across Persia to link the Indian and Russian systems. The entire length of line to be constructed, according to this plan, from the last station on the present Indian system to the nearest Russian station in the Caucasus is 1994 miles.

Occurrences in Macedonia are palpable evidence of the continuity of relationship between the unfortunate measures of the Congress of Berlin, in 1878, and the turmoil which has convulsed the Balkan peninsula during the past few years. The Treaty of Berlin created the Macedonian question by annulling the incorporation of the greater part of the territory with Bulgaria, and restoring it to Turkish misrule. For the Christians of Macedonia never relinquished the hope which had been once evoked, and the prospect of the liberty of their brethren in Bulgaria, Servia, and Greece made their own situation more intolerable. Macedonia became constantly restless, and nationalistic societies in the neighboring countries contributed continually to this condition of ferment by providing the material means for insurrection. From time to time, as the patience of the Turkish military element became exhausted, their exasperation found vent in savage reprisals. The Macedonian disturbances at length became so excessive that they could not be ignored by the powers. The Sultan had promised by the Treaty of Berlin to reform the administration of Macedonia and had treated this engagement with the same indifference as many other plausible assurances of the same nature. The powers insisted upon the application of the so-called Mürzsteg program in 1903, by which the three districts Salonica, Monastir, and Kossovo were placed under the supervision of Austrian and Russian civil agents, and the Macedonian gendarmerie was reorganized under the able direction of an Italian officer, General De Giorgis. The Macedonian problem was unfortunately not based upon an ideally simple contest where the forces of tyranny are ranged on one side and the down-trodden on the other. It was complicated by the heterogeneous character of the Christian population, which is composed of Greek, Serbian, and Bulgarian elements. The Turkish sovereignty partly concealed the really more essential antagonism, which grew out of the rivalry of the Christian nationalities, particularly the Greeks and Bulgarians. The Greeks claim to be the élite of the Balkan peninsula. Proud of the matchless distinction of their pedigree and their exalted traditions of civilization, they regard with a certain measure of contempt the Bulgarians as descendants of "barbarians." The conviction had prematurely taken root in Macedonia, that in the final partition of the territory, political boundaries would be traced along the lines of demarcation of the different nationalities. It was a policy of timely foresight, therefore, for each nationality to endeavor to make its allotment as large as possible in anticipation by exterminating as much as possible of the rival populations. Bands were accordingly organized in each of the three countries that had prospects in the partition of Macedonia for the purpose of clearing the soil of its competitors. This irregular warfare was conducted with the greatest ferocity by the Greeks and Bulgarians. In fact, the second Balkan war was simply the resumption of a natural and inevitable condition of hostilities after a brief interlude for ousting the Turks, incidentally a common aim.

The Mürzsteg program was inadequate, chiefly, no doubt, on account of the incessant activity of the guerilla bands. At the time of the meeting of Edward VII and Nicholas II at Reval in the summer of 1908, a more radical scheme was approved. But before it could be put into execution, the most unexpected, the most remarkable, revolution had occurred in Turkey almost without bloodshed, so that the foreign officials were withdrawn from Macedonia with the supreme confidence of the powers that an era of domestic reconciliation and brotherhood had dawned for the distracted Turkish Empire.

The progressive partition of the Turkish dominions had undoubtedly wounded deeply the pride of the more thoughtful Turks. The conviction gained ground that a thorough transformation of Turkey from within was the only means to counteract the obvious retrogression, and there were not lacking men of integrity, with faith in their country, who believed that this was possible. Besides, there was an increasing army of malcontents and exiles, victims of the Hamidian despotism. By the combination of such elements, the Young Turk party was constituted with a program of liberal nationalism. Its directing organ was the Committee of Union and Progress. This committee carried on a propaganda with the utmost dexterity and success. Many of the leading officers in Salonica and throughout Macedonia became distinguished members of the committee; among them Enver Bey. The prospect of another curtailment in the extent of the Sultan's government by the practical independence of Macedonia spurred on the Young Turks to prompt measures. A revolution was inaugurated in Macedonia, and two days later it culminated in Constantinople with the proclamation by Abdul-Hamid of the restoration of the constitution of 1876. The world was astounded, and public opinion, naively prone to be captivated by the paradoxical, was ready to predict the speedy regeneration of the Turkish nation. It is a fact, however, that the Young Turkish leaders gave proof at this time of great capacity and moderation. Kiamil Pasha, reputed an Anglophile, was appointed Grand Vizier, August 6, and intrusted by Abdul-Hamid with the formation of the first constitutional cabinet.

Unfortunately, the new order was assailed immediately with difficulties from without and within. Despite universal expressions of sympathy, the prospective heirs of the Ottoman Empire really regarded this act of rejuvenation as an unwarranted postponement of the demise of the incumbent of their rightful inheritance. They undertook with cynical haste to forestall a regeneration which might postpone the acquisition of their spoil indefinitely. The election of members for the approaching Turkish parliament would be a formal act indicative of organic union with the empire. It was very likely regarded somewhat as a test of corporate unity in the border, or doubtful, provinces. The summoning of the first parliament probably accelerated, for this reason, a series of proclamations declaring specifically the exclusion from the Turkish Empire of some of these provinces, which must have been a bitter disappointment to the Young Turks. Thus, on October 5, Ferdinand, Prince of Bulgaria, assumed the title king, or tsar, renouncing thereby the suzerainty of the Sultan, and two days later the Cretans repudiated their connection with Turkey, and Francis Joseph solemnly proclaimed the definitive annexation of Bosnia and Herzegovina to the Hapsburg dominions.

The momentous act of the Austro-Hungarian government in annexing the provinces which it had been administering since 1878 was chiefly due to the spirited policy of Count Aerenthal, at that time Minister of Foreign Affairs. The Young Turks accepted these repulses in a spirit of moderation, and entered into negotiations with their neighbors on the north with a view to the peaceful adjustment of the situation. An agreement was finally concluded with Austria-Hungary in January, 1909, the principal features of which were the Sultan's recognition of the Austrian annexation of Bosnia and Herzegovina, the renunciation by the Austrian government of all rights in the sanjak of Novi Bazar, the narrow strip of territory which at that time separated Serbia and Montenegro, and payment by Austria-Hungary of a financial indemnity for the property of the Turkish state in the annexed provinces.

The bitter resentment of Serbia offered the most serious problem arising from the annexation of Bosnia and Herzegovina. We have already commented upon the rise of the Greater Serbian idea. It is not difficult to appreciate the intense feeling of injustice which rankled in the hearts of the Serbians. Only about one-third of their particular branch of the Slav race was included in the Kingdom of Serbia. The transfer of the administration of Bosnia and Herzegovina into the hands of the Hapsburg government by the Congress of Berlin had been a cruel disappointment, but the measure was ostensibly temporary. Now the definitive annexation threatened to perpetuate for all time the politically divided condition of the Serbian people. In annexing the provinces the Dual Monarchy had usurped a prerogative which belonged to the signatory powers of the Treaty of Berlin. The Serbian Skupschtina, or parliament, passed a resolution demanding the autonomy of Bosnia and Herzegovina under the protection of the powers, and for themselves a land connection with Montenegro, which would give Serbia an outlet to the Adriatic Sea guaranteeing her commercial independence. But the government of Austria-Hungary asserted that their annexation of Bosnia and Herzegovina was an act which concerned the nominal

German Empress as a Mediæval Queen.

William II as King of Jerusalem.

When the Emperor and Empress visited Jerusalem in 1898 the latter chose, on the Mount of ives, the site for a German hospital, named the Augusta Victoria, in the courtyard of which these statues by Moritz Wolff.

proprietor of the provinces and themselves alone, and since Turkey had already concurred in it, the incident was closed. The tension became extreme during the winter months and military preparations were made by both parties. But Serbia had no prospect of success in the face of the determined attitude of her powerful opponent, unless she could count on the assistance of Russia. The empire of the Tsar had not recovered from the calamities of the war with Japan and the disorganizing effects of the revolution. The government hesitated before the likelihood of a European war. Germany exerted the decisive influence at St. Petersburg, and it is said that her representations were peremptory in tone. Serbia was constrained to await the decision of the powers with regard to Bosnia and Herzegovina, which was a formula intended to spare Serbia's feelings by disguising what was really submission, since the powers were not disposed to contest the annexation. On March 31, 1909, the Serbian minister in Vienna, on instructions of the Serbian government, made the following declaration to the Austro-Hungarian government:

"Serbia recognizes that the *fait accompli* regarding Bosnia has not affected her rights, and consequently she will conform to the decisions that the powers may take in conformity with article 25 of the Treaty of Berlin. In deference to the advice of the Great Powers, Serbia undertakes to renounce from now onwards the attitude of protest and opposition which she has adopted with regard to the annexation since last autumn. She undertakes, moreover, to modify the direction of her policy with regard to Austria-Hungary and to live in future on good neighborly terms with the latter."

The crisis arising out of the annexation of Bosnia and Herzegovina by the Dual Monarchy resulted in a victory for German diplomacy. The prompt loyalty displayed by the ally was reflected in enthusiastic expressions of gratitude in the Austro-Hungarian Monarchy, and the moral ties between the two empires were undoubtedly strengthened. We may note in Chancellor von Bülow's report of the affair before the Reichstag the following passage:

"The Serbian demands are a perilous adventure; and while it would be intolerable to think of European peace being threatened on account of Serbia, it would be absurd to pretend that Austria-Hungary or Turkey should yield to the unreasonable aspirations of Serbia. No war at all, far less a European war, should spring from such aspirations."

It is scarcely necessary to add here that the Serbians did not resign their bitterness and sense of injury.

Germany must have regarded with grave anxiety the subversion of the old order of things in the Ottoman Empire in 1908. By cultivating the situation for thirty years with diligence and tact, the Germans had acquired a position of ascendancy, to which a revolution could assuredly add nothing; while, on the other hand, the attitude of the leaders in the new government was at first distinctly favorable to the western powers.

There was a counter-revolution early in 1909; but it was of very short duration. On April 14, mutinous soldiers in Constantinople seized the parliament house, and forced the Grand Vizier to resign; but the Committee of Union and Progress sent a force of 25,000 men from Salonica, under Mohammed Shevket, who occupied the capital after some severe street fighting, April 25, deposed Abdul-Hamid, and elevated to the sultanate his brother as Mohammed V, after he had been confined thirty years as a prisoner.

The Young Turkish revolution had been inspired by a passionate desire to regenerate the Ottoman Empire by the introduction of parliamentary institutions. Liberalism and nationalism were, therefore, intimately associated in the program of the Young Turks. The movement had been impelled by the determination to counteract the threatening disintegration of the empire. The Young Turks were, therefore, opposed on principle to all separatist tendencies within the empire, including demands for the special treatment of particular nationalities or provinces. The fulfilment of their purpose required that the aspirations of separate nationalities should be sacrificed to the liberty and prosperity of the empire as a whole. This attitude of the new element in power probably facilitated the recovery of German prestige and influence, because Germany had never embarrassed or annoyed the Turkish government by urging special reforms, and no interested motive made her the advocate of any of the Christian nationalities in the empire. Moreover, Germany had been represented in Constantinople, since 1897, by a diplomat of rare ability, Baron Marschall von Bieberstein. He performed a difficult service for the Fatherland, of inestimable value; but he must have encountered a situation that taxed his resourcefulness to the uttermost, when the war broke out with Italy in the autumn of 1911.

We must be impressed by the heartless waywardness of fate, when we reflect that during thirty years under the despotism of Abdul-Hamid the peaceful relations of the Ottoman Empire were disturbed by only one foreign war, and that of brief duration, whilst the generous impulses of the Young Turks have exposed Turkey to many acts of unfriendliness, constant turmoil, and three foreign wars within a period of seven years. It was not enough that one of Germany's allies should have rudely humiliated the new government of Turkey two months after its establishment. Before Europe had recovered from the anxious tension of the Moroccan controversy, Italy, the other ally of Germany, delivered at Constantinople her ultimatum, demanding, in terms that precluded procrastination, the cession to her of Tripoli, a natural goal of Italian policy.

Tripoli is a territory of possibly 400,000 square miles, lying directly opposite the southern extremity of Italy, between the eastern confines of the French colonial empire in Tunis and the western boundary of the British sphere of control in Egypt. It was sparsely inhabited; but its importance in ancient times attested the existence of a considerable area of fertile soil. The possession of this nearby territory would afford precisely the opportunity which Italy required, a field for colonial expansion, where the energy of her surplus population could be profitably employed, and an added position of strategic importance in the Mediterranean. The aspirations of Italy regarding this territory had long been recognized by France and Great Britain.

We naturally suspect that the remarkable events which have occurred since 1911 in such rapid succession, even overlapping one another in a chronological sense, are the consequence of a closely related group of causes, perhaps the counterplay of the same diplomatic forces. Italy launched her enterprise as soon as the probability of a French protectorate in Morocco left Tripoli as apparently the only territory on the Mediterranean open to fresh projects of colonization. But some other circumstances must be considered in canvassing the possible causes of the situation, and they suggest that larger motives than Italy's cupidity were involved in the venture.

With Tripoli in her hands, it seemed that Italy might be regarded as the key to the Mediterranean; and it might therefore be assumed that such a position would have been considered by Italy's partners in the Triple Alliance as a decisive advantage to themselves, and that Italy invaded Tripoli with their concurrence, if not at their suggestion. But on the other hand, the close union with Turkey was the

The Sultan of Turkey, Mohammed V, reading the prayer at the sacrifice of Qurbān Bairām, when a ram is slain to commemorate the ransom of Ishmael with a ram. The ceremony taking place at the Imperial Palace, Constantinople.

very corner-stone upon which the most attractive part of German aspirations for expansion had been erected. Italy's aggression must have imperilled the influence of her ally in Constantinople. It is a brilliant achievement of German diplomacy to have recovered its controlling position in Turkey. But the fact that German diplomacy weathered the gale is no proof that it provoked it. Besides, for Italy, membership in the Triple Alliance had been due solely to a calculating, unemotional policy. Attachment to it waned as the material advantages appeared less obvious, and the increasing feeling of rivalry between Italy and Austria was a positive danger to the continuation of the compact. Austria's more favorable naval situation in the Adriatic, the best harbors of which are nearly all on the eastern, or Austrian side, and her imputed inclination to penetrate farther into the Balkans rendered Italy ever watchful and suspicious. It was, undoubtedly, another triumph for German diplomacy to have maintained outward harmony so long between these uncongenial allies.

By contemplating these circumstances, and recalling that Germany's Moroccan enterprise was regarded in France and England as an aggressive operation aimed at their *entente*, we are naturally led to the conjecture that the western powers suggested Italy's abrupt action in Tripoli as a prompt counter-thrust, calculating that it would embarrass hopelessly German foreign policy and break up the Triple Alliance. It will appear from all this how delicate and all-important must have been the diplomatic task of the German representative at Constantinople. And Germany managed in some way to reconcile her position as ally of Italy with that of friend of Turkey.

It is foreign to the purpose of this volume to relate the events of the campaign in Tripoli. We need only recollect that Turkey resisted this further operation of dismemberment with unexpected tenacity, and that to reduce her to yielding Tripoli, Italy occupied a number of the Aegean islands. But before the final settlement between Turkey and Italy, a more formidable storm of calamity had broken over the unhappy Ottoman Empire.

The establishment of the constitution by the Young Turks was followed by an effusion of goodwill throughout the empire. Races which had been inveterate enemies embraced and fraternized. But the disillusionment began when the Christian populations lapsed into their congenial state of dissension. The spread of liberal ideals intensified the spirit of nationality in Turkey. Intolerance was engrafted as a branch on the tree of liberty; and this was just as true of the Christians as of the Turks. The Young Turkish movement for the redemption of the Ottoman Empire was largely a failure because the nationalities comprised within the empire were too distinct, racial barriers were too abrupt, for the development of a general, comprehensive patriotism ardent enough to fuse the feelings of the particular peoples into an enthusiastic consciousness of their common country and community of interests. Accordingly, the discontent in European Turkey again became threatening.

An insurrection of the Malissori, a Christian tribe of northern Albania, in May, 1912, was followed by a mutiny of the Turkish troops at Monastir in June, and the fall of the Young Turkish ministry. Ghazi Mukhtar Pasha became Grand Vizier, and Nazim Pasha, Minister of War, July 21. The Turkish troops returned to their allegiance, but the spirit of insurrection spread from Albania to Macedonia, where it pervaded the three Christian populations. The Turkish soldiers committed revolting atrocities, a massacre of Bulgarians at Katchana and vicinity in August, and of Serbs at Berane and Sienitza. It became very difficult for the governments of the Christian states to resist the popular enthusiasm for intervention to rescue their brethren in Macedonia. The pressure was especially strong in Bulgaria, whither thousands of Macedonians of their own nationality had taken refuge. The Balkan powers at length became convinced that the time had come for the fulfilment of national liberation and the unity of their respective peoples.

The Balkan League seems to have been conceived in the first place as a union of all the Balkan powers, including Turkey, to support, ostensibly at least, the doctrine of the "Balkans for the Balkans." This earlier plan may very likely have been a suggestion of the German ambassador in Constantinople, in the hope of seeing the Balkan states present an undivided front against Russian interference, without compromising Germanic peaceful penetration. But this project, if ever seriously considered, was soon abandoned. It lacked a vital, compelling motive.

The league for aggressive action against Turkey was probably inspired by Baron von Hartwig, the Russian ambassador at Belgrade, and Venizelos, the very able Greek Prime Minister. It was based officially on a treaty between Serbia and Bulgaria, signed February 29, 1912, and another between Greece and Bulgaria, signed May 16, 1912. The former of these treaties consisted of two parts, one part suitable for publication, establishing a purely defensive alliance, and a secret part, which is most important, since it contains an agreement relative to the division of territory between Serbia and Bulgaria in the event of a victorious war of aggression against Turkey, and reveals a connection of Russia with this proposed movement by naming the Tsar as umpire in all questions arising between the allies.

Relations between the new allies and the Ottoman Empire became more and more strained, so that the former commenced mobilization, September 30, and the latter, the next day. Austria-Hungary announced as the basis of her policy in this crisis the maintenance of the status quo, and, ostensibly, non-intervention. Austria and Russia issued a declaration, October 8, as mandatories of the Great Powers, "condemning any measures susceptible of a breach of the peace," announcing that the powers would take in hand the necessary reforms according to the Treaty of Berlin, and adding the solemn admonition that in the event of hostilities the powers would "not admit at the close of the war any modification of the territorial status quo of European Turkey." But the Balkan states were determined to make the Treaty of Berlin as obsolete as it had been stupid, and the formal declaration of the powers as to the immutability of Turkish boundaries was treated with the indifference which the outcome proved that it deserved. The conduct of Austria and Russia had not been uniformly of such a character as to invest their warning with unqualified respect, since the former had herself manifestly slighted the authority of the powers in annexing Bosnia and Herzegovina, and the latter was apparently entirely cognizant of the aggressive intentions of the Balkan states, as the secret treaty seems to indicate. It might be urged in this connection that since duplicity is assumed in the game of Balkan politics, straightforwardness would have the effect of more subtle artifice, and might be regarded, therefore, with serious moral misgivings.

Montenegro, the irrepressible, pugnacious bull-dog of the Balkans, instinctively impelled against its congenital foe, slipped the leash and declared war, October 8, before negotiations had run their normal course, perplexing, no doubt, the counsels of the allies, and causing uneasiness throughout Europe. Thus Montenegro precipitated the Balkan War with its momentous consequences. The allies followed her initiative, October 18, each state throwing its forces into the

House of Parliament, Constantinople.

The Russian Embassy at Constantinople.

territories adjacent to its own boundaries, chiefly such as were inhabited by people of the same nationality.

The aggregate war forces of the Balkan allied states amounted to 790,000 men, and those of the Ottoman Empire were supposedly about 700,000. German officers had reorganized the Turkish military establishment, as has been pointed out, and held important commissions in the armies. Austrian officers had in large measure performed the same service for the Bulgarians, whilst a French military commission had supervised the development of the military organization in Greece. The Greeks and Serbs brought the Creusot French field-artillery into action against the Krupp German guns of the Turks. These circumstances were not absent from the minds of the military experts, who followed the course of the operations with eager attention.

A feeling of relief was experienced by all those who felt sincere concern for the good name of Italy, when she straightway concluded peace with Turkey without profiting by the unfortunate position of her adversary to augment her demands, October 15. To palliate her submission, Turkey retained the privilege of a representative in Tripoli to protect the interests of the Mohammedans. Italy pledged herself to restore the islands which she had occupied in the Aegean Sea, provided Turkey would grant them autonomy; but the rapid progress of the conflict between Turkey and Greece made their retention by Italy expedient until their eventual destiny should be decided in the final settlement after the Balkan war.

The Serbs quickly overran the greater part of Macedonia and northern Albania as far as the coast of the Adriatic Sea. The Greeks entered Salonica, November 8, and the Bulgarians arriving before the city a little later exhibited ominous signs of dissatisfaction. The more arduous field of operations in the direction of the Turkish capital fell to the portion of the Bulgarians, where their swift, overwhelming successes astonished Europe. After their defeats of Kirk Kilisse, October 22 and 23, and of Lule Burgas, October 28, the Turks fell back to the Tchataldja lines, the outer defenses of Constantinople, leaving a strong garrison in Adrianople. Their disheartening misfortunes caused a revulsion of feeling, the return of the Young Turks to power, and the adoption of a pacific attitude. Kiamil Pasha was made Grand Vizier, and an armistice was signed, December 3, to which Greece, however, was not a party.

In the mean time Albanian delegates had assembled at Valona, November 28, and had set up a provisional government under the presidency of Ismail Kemal Bey.

The convention of envoys from the Balkan states and Turkey for elaborating terms of peace, and a conference of the ambassadors of the Great Powers for coördinating their own views began their sessions coincidently in London, December 16. The vicissitudes of war and negotiation are reflected in the petulant course of Austrian remonstrances, to which the powers displayed undeniable proofs of the utmost consideration. The Dual Monarchy must have recognized the impossibility of upholding the territorial integrity of European Turkey after Kirk Kilisse and Lule Burgas. For a time the Austrians insisted that the sanjak of Novi-Bazar must be respected by the belligerents, until the occupation by the Serbians of all the territory to the south of it deprived the reservation of the sanjak of its former significance.

But the march of Serbia across northern Albania to the Adriatic Sea, and her avowed intention of extending her boundaries to the coast led to partial mobilization in Austria-Hungary, which brought her forces under arms up to about 1,000,000 men. Russia responded to this measure by the concentration of unusual forces on the frontier of Galicia.

In November, the Archduke Francis Ferdinand, heir presumptive of the Hapsburg realm, had an interview with William II in Berlin to discuss the subject of the military coöperation of the two empires in case war with Russia should follow, and the nervousness in Austria-Hungary showed itself in a financial panic.

The idea of an independent Albania, as accepted by the conference of ambassadors in London, December 20, had been devised long before in the consultations of Austria-Hungary and Italy regarding their common policy in the possible event of a dissolution of the Ottoman Empire. It was a contrivance for insuring the stability of the existing equilibrium of power in the Adriatic Sea.

The Albanians are the descendants of the ancient Illyrians, who were at one time famous as pirates, and they retain to-day the turbulent characteristics of their forefathers. The hardy mountaineers of Albania were the last of the Balkan populations to submit to the Turkish invaders; but most of them later became Mohammedans, and furnished a vigorous element to the armies of the Ottoman Empire. In this respect Albania is unique among the states which have been erected on fragments of the former Ottoman territory in Europe. The present Albanians preserve a tribal organization which engenders animosity and numerous feuds. Robbery and brigandage among them pass over by scarcely perceptible degrees into heroic and patriotic warfare, as in the Greece of Homeric days. The Albanians are, unfortunately, destitute of some of the necessary requisites for political independence on a national basis, such as a literature, and common moral ideals. The delimitation of the new principality was the thorniest problem with which the conference of ambassadors in London had to deal.

In the first place, Serbia expected at least enough of Albania to form a continuous extension of her territory to the Adriatic Sea. Serbia maintained that an opening to the sea was indispensable for her independence, on account of Austria's unfriendly tariff policy which aimed to make her a commercial vassal. But Austria-Hungary was inflexibly opposed to any Serbian maritime outlet whatsoever. She even refused her assent to the suggestion of a narrow strip of Serbian territory, a "corridor," as it was called, extending from the central mass of Serbian territory to some sea-port on the Adriatic. Austria-Hungary appeared to have singled out Serbia as the special victim of her displeasure with the determination of thwarting her purposes at every turn. She regarded Serbia as an outlying province of a Panslavist Russia, and a Serbian port as an available naval base for the same power.

The powers made their decision respecting Albania in compliance with the wishes of Austria-Hungary. But it was a weighty task to restrain the impetuous Serbians in their eager course; and it was probably due to the urgent representations of Russia in the cause of concord that her Slav protégé sorrowfully renounced the most coveted goal of her exertions. Eventually the powers confirmed the occupation of the eastern part of Albania by Serbia and Greece, limiting the new principality to the western part of the country, which faces the Adriatic Sea.

The great obstacle to the conclusion of peace between the Turks and allies was the determination of the former not to give up Adrianople, a place which was hallowed by Ottoman traditions of peculiar sanctity, and was regarded as strategically indispensable for the defense of Constantinople. It is necessary to point out that, although the Turks had everywhere been defeated in the open, their three principal strongholds in Europe, Adrianople, Yanina, and Scutari, had not fallen. Besides, their forces before Constantinople had been constantly strengthened by reinforcements from

Asia. Accordingly, when Kiamil Pasha was about to yield to the demands of the allies, Enver Bey and his companions of the war party forced their way into the council chamber, and compelled Kiamil Pasha to sign his resignation, January 23. During this violent incident, Nazim Pasha, the unfortunate commander-in-chief who had lost the battles at Kirk Kilisse and Lule Burgas, was killed. Mohammed Shevket Pasha, who became Grand Vizier, will be recalled as the general who put down the counter-revolution in 1909. He was probably the most able and trustworthy statesman in the empire, though of Arabic, not Turkish, nationality.

As a consequence of the sudden turn of affairs at Constantinople, the peace convention in London dissolved, February 1, hostilities were at once resumed; but Yanina surrendered to the Greeks, March 9, and Adrianople to the Bulgarians, March 28. The Montenegrins had been besieging Scutari since the early part of the war. The second armistice between the other combatants, April 20, did not check the operations before Scutari. At the demand of Austria-Hungary, the powers had decided that Scutari must form part of Albania; and for weeks Scutari and Montenegro were the storm center. Europe was kept in a constant state of nervous excitement by the obstinate insistence of Austria-Hungary, and the inflexible perseverance with which the Montenegrins pressed the siege. The average observer of events might have felt that after the collapse of several unreasonable expectations, Austria-Hungary had seized upon the case of Scutari in a spirit of irritated perversity, determined to win her point, as a satisfaction to disappointed pride. It is true that Scutari is the only city worthy the title within the proposed boundaries of Albania. It is the center, moreover, of the Roman Catholic tribes of northern Albania, who were under the protection of Austria-Hungary. Austria protested emphatically that Albania was impossible without Scutari; but experience has yet to prove that Albania is a possibility with Scutari.

The Great Powers invited Montenegro to desist from besieging Scutari, on April 1, since its capture could bring them no profit. But Montenegro continued to besiege Scutari in open defiance of the Great Powers, although an international squadron proceeded to blockade the two villages of Montenegro's sea-board. Finally, Essad Pasha, the commander in Scutari, signed the capitulation, April 22, and, eager to exchange the rôle of Turkish general for that of prince of Albania, he issued a proclamation of Albanian independence a few days later.

But King Nicholas of Montenegro, assured, no doubt, by Russia, that the situation permitted no alternative, submitted to the most painful renunciation which his people has ever endured. The possession of Scutari had been a hereditary aspiration of the Black Mountain folk, to which they had devoted superhuman exertions during the war that was just closing. On May 5, the king sent a telegram to Sir Edward Grey, placing Scutari in the hands of the powers. With sullen resentment in their hearts, the Montenegrin forces withdrew from Scutari, nourishing a feeling of intensified bitterness against the harsh Hapsburg neighbor.

The disagreement between the Balkan allies regarding the new boundaries pressed closely upon the heels of the events which have been described, keeping the diplomacy of Europe constantly distracted. A Greco-Bulgarian boundary commission separated, May 9, after having deliberated for two months, unable to arrive at any agreement, and the negotiations between Serbia and Bulgaria were equally unsuccessful. The basis of the controversy was substantially the demand of Bulgaria that the new boundaries should be adjusted in harmony with the agreement of 1912, without any modification of it to compensate Serbia and Greece for the exclusion of the territory of Albania from their portions.

A second Balkan peace conference assembled in London, May 20, at the initiative of Sir Edward Grey, for the arduous purpose of reducing to peaceful equilibrium by calm, dispassionate deliberation the turmoil of conflicting hopes and ambitions. This conference accomplished nothing that was permanent, although it seemed a substantial achievement when a treaty was signed between Turkey and the Balkan states, May 30, which established the so-called Enos-Midia boundary-line for European Turkey, leaving Adrianople in the possession of Bulgaria. But the situation between the allies themselves rapidly grew more threatening, and the conference separated, June 9, helpless to check the coming storm.

An urgent telegram from the Tsar, who had been constituted mediator by the treaty of alliance, deprecating the fratricidal war that impended was of no avail. The insistent demand of Roumania for a further rectification of her boundary with Bulgaria (she had already received Silistria) introduced a new element into the situation. With headstrong pertinacity, Bulgaria persevered in a haughty, uncompromising attitude, and even took the offensive against Greece and Serbia, June 30. But Roumania intervened, July 9, and Bulgaria could not detach any of her forces to resist the advance of the Roumanian army. The Turks seized the opportunity to resume hostile activity, and, under the leadership of Enver Pasha, they reoccupied Adrianople, July 31. The essential feature of the second Balkan War was the disastrous course of Bulgarian operations, and the rapid collapse of her military power. The Bulgarians were forced to ask for an armistice, July 31, and delegates from the belligerent states immediately assembled at Bucharest, where a treaty of peace was signed, August 10. The changes in area and population brought about by the two wars, as sanctioned by the Treaty of Bucharest, may be briefly tabulated, as follows:

States	Before Balkan Wars Area in sq. miles	Population	By Treaty of Bucharest Area in sq. miles	Population
Turkey in Europe..	65,350	6,130,200	10,882	1,891,000
Greece...........	25,014	2,975,953	41,933	4,363,000
Montenegro.......	3,474	about 250,000	5,603	516,000
Serbia............	18,650	2,911,701	33,891	4,547,992
Bulgaria..........	33,647	4,337,516	43,305	4,752,000
Roumania........	50,720	7,230,418	53,489	7,516,418
Albania..........................			11,000	825,000

Note: The populations before the wars are taken from the most recent census in each country.

An agreement was concluded between Bulgaria and the Ottoman Empire, September 18, which altered the "Enos-Midia" boundary in such a way as to secure for the Turks about one hundred square miles of additional territory in Europe, including the much-disputed Adrianople.

An international commission of control for Albania was set up by the powers. It assembled at Valona the end of September, and selected William Frederick Henry, Prince of Wied, as ruler of the new principality of Albania, November 23, 1913.

To fully appreciate the significance of the vicissitudes of power in the Balkans, we must observe once more the natural trade-routes. In connection with the Danube, it is important to note that its upper course is united with the Main and Rhine, and the other important waterways of Germany, by means of canals, so that it affords the cheapest means of transportation from the great industrial centers, as Westphalia, to the Black Sea, and thence by the Bosphorus and Dardanelles to the Aegean Sea. But nature has prepared an alternative for the lower Danube, a quite direct route

Francis Ferdinand with his wife and children.

Francis Joseph with his wife and children.

from a point just below Belgrade along the valleys of the Morava and Vardar Rivers to Salonica, and a railway follows this natural highway. Moreover, a natural route proceeds from Sarajevo in Bosnia in a southeasterly direction to Uskub, where it joins the Vardar route. The Austro-Hungarian government had demanded a concession for a railway line along this route a short time before the Young Turkish revolution in the Ottoman Empire. These two routes from the north, traversing Serbia and Bosnia respectively, explain the great importance of Salonica.

Austria-Hungary had imposed her will upon the counsels of the powers throughout the Balkan conflicts in the specific cases where she considered that her vital interests were affected, and had frustrated the most cherished designs of the neighboring Slavic states. Yet the crisis left her dispirited and humiliated, with diminished prestige and security; and as an added aggravation, she had no rational, concrete grounds for complaint. She had advocated ostensibly the doctrine of the "Balkans for the Balkans," asserting that the Balkan states should be left to settle their own destiny. This they had presumably accomplished by appealing to the verdict of the sword, and with results which were satisfactory to the majority of them, at least. Let us regard the situation in detail from the Austro-Hungarian point of view. Serbia and Montenegro had absorbed the sanjak of Novi-Bazar and brought their boundaries together. They were, henceforth, virtually a single state as regards their foreign relations, and were naturally animated with a sentiment of great bitterness towards Austria-Hungary. They lay athwart the roads to Salonica, abruptly shutting off the enticing prospect of Macedonia and an outlet onto the Aegean. Moreover, the practical union with Montenegro might compensate Serbia for her exclusion from a doorway to the sea in Albania. For in one of

Montenegro's sea-port villages, Serbia's future emporium and naval base might conceivably be created. Austria-Hungary had mortally offended her neighbors, and gratuitously too, which is surely a very unsagacious policy. Then, the defection of Roumania must be recorded, in associating herself with the unfriendly states for the spoliation of Bulgaria, who was regarded as almost an ally by the Dual Monarchy. In short, a combination of powers, Roumania, Serbia, Montenegro, and Greece, whose policy was incompatible with friendship for Austria-Hungary, was supreme in the Balkan peninsula, and Bulgaria and Turkey were apparently out of account.

It was immediately rumored that Austria-Hungary would demand the submission of the Treaty of Bucharest to a revision by the Great Powers on the ground that it did not provide a stable equilibrium, and, therefore, did not offer guarantees for permanent peace in the Balkans. It has since been shown, that on August 9, Austria-Hungary informed Italy of her intention of delivering an ultimatum to Serbia substantially identical with the memorable message actually despatched nearly one year later. Austria-Hungary regarded the proposed step as virtually a defensive measure, which would, therefore, involve Italy's obligation to render assistance; but Giolitti, who was at that time prime minister of Italy did not view the proposal in the same light. Austria-Hungary undoubtedly approached her greater ally with this same proposition, and it was doubtless the pacific attitude of the Kaiser's government which put off the outbreak of the world-war for another year.

We must never lose sight of Germany's economic interest in the Ottoman Empire in its relation to the political situation. Serbia alone stood between the Teutonic empires and a continuous field for unlimited commercial and financial expansion in friendly or economically tributary countries traversed by the Oriental, Anatolian, and Bagdad Railways. This view is strikingly illustrated by an article in the *Frankfurt Gazette*, January 7, 1914, as the following extracts will show:

"The states lying between the eastern border of Hungary and Asia Minor have indeed no choice; they must be the friends and allies of the Triple Alliance, or they must reckon with the unflinching hostility of the Triple Alliance in any conflict which threatens their independence.......... Austria too has no choice. Either the countries on the Lower Danube must be her friends or she must exterminate them..........When Bismarck said that the whole Balkan peninsula was not worth the bones of a single Pomeranian grenadier, he could not have foreseen that this territory would one day become so essential a route for German imports that we should not, if necessary, shrink from a conflict with Russia to maintain our freedom of trade there."

But the disasters of the German-trained Turkish armies had lowered Germany's military prestige, and the results of the Balkan conflicts profoundly disturbed the military calculations of both the Teutonic powers. Intimate relations with Turkey had been the essential feature of Germany's policy in the Near East. Roumania had been regarded as a natural ally on account of her resentment for the loss of Bessarabia to Russia, and Bulgaria's friendship seemed to be assured by the bonds of self-interest. The friendly elements in the Balkan peninsula had apparently more than counterbalanced the forces of possibly hostile states, so that, in the event of a European war, Austria-Hungary would have been able to employ almost all her military forces against Russia. Thus the military situation made the Triple Alliance, and more particularly the Teutonic powers, seem quite secure until the eve of the collapse of Turkish power in Europe. Now, however, Turkey was apparently overwhelmed, Bulgaria was prostrate, and Roumania's conduct permitted no hopes of friendly co-operation.

It was plain that for Germany to retain the same favorable situation of military security, a material increase in her military establishment would be necessary.

The text of a proposed new army bill was first published March 28, 1913, and it called for an increase of 4,000 officers, 15,000 non-commissioned officers, and 117,000 corporals and privates, and 27,000 horses in the standing army, which would bring the aggregate forces on a peace footing up to 870,000 men. There had been 280,000 annual recruits up to this time; henceforth there would be 340,000. One noteworthy feature of the new bill was its aviation section, providing for a station for air-ships and two squadrons of four air-ships each, together with one headquarters and six subsidiary stations for a total of fifty aeroplanes. In all, 1542 men would be attached to this branch of the service.

This vast addition to the strength of the army would involve a corresponding increase of about $45,000,000 in the annually recurring expenses, whilst a non-recurring expenditure of about $252,000,000 would be required for the requisite barracks and material. Accordingly, a finance bill accompanied the measure for augmenting the army, providing a special tax on all private fortunes from a minimum valuation of 30,000 marks (about $7200), the rate increasing gradually to 1½%, and a tax on all incomes from 5000 marks (about $1200) which are not derived from property which would be subject to the tax on capital, the rate increasing gradually from 1% to 8%. These extraordinary taxes, called the *Wehrbeitrag*, or contribution for defense, were to be paid in three equal, annual instalments.

In his speech justifying these measures before the Reichstag, the Chancellor explained that events in the Balkans had

wife were assassinated, June 28, 1914

shifted the balance of power so that in the event of war Germany might have to protect herself against several enemies on extensive frontiers poorly defended by nature. He thought that there would probably be no European war in which Germany would not be involved, and it would be a struggle for existence. No great power desired war; but nobody could be sure that it would not break out at any time; for the moderation and feeling of responsibility of the powers had alone prevented the strained relations between Austria-Hungary and Russia from developing into war. Germany enjoyed very friendly relations with Russia, and antagonism was not likely to arise of itself in her relations with her great Slav neighbor. But the Panslavist movement had received a powerful stimulus from the victories of the Slavs in the Balkans, and they had to consider this in planning for the future. The influence of public opinion in determining the policy of nations had increased very much, and often a minority element was the most violent and obtrusive. But many of the quieter elements in France believed that their army was equal or superior to the German, and based hopes on their alliance with Russia, and perhaps the *entente* with England. Russia, moreover, was reorganizing her military establishment on an unparalleled scale.

The army and finance bills were introduced into the Reichstag, April 7, and passed their third reading together, June 30. It is a significant fact that only Poles and Socialists voted against the army bill, and even some of the Socialists voted for the finance bill, approving the property tax on principle.

In view of the alacrity with which this unforeseen and heavy burden was accepted, one is at a loss whether to marvel most at the spirit of unity animating, in this supposed situation of great national necessity, the upper and middle classes in Germany, or at the dexterity with which the government manipulates the parties and controls important influences moulding public opinion. It is probably safe to say that a military measure involving such sacrifices could not have been passed in any other constitutional country in the world without the incentive of a more imminent, or visible, danger.

It is true that equally radical proposals were immediately suggested in France, but solely as a consequence of these extraordinary German measures. The feeling was expressed quite generally in France that the security of the country was menaced by the sudden increase of the German military establishment, and that they must return to the term of military service for three years to counter-balance the very great element of German military superiority due to the greater population of the Teutonic empire. There was a rupture in the ranks of the Radicals on the question of the period of service. The Briand ministry was defeated, and M. Barthou was entrusted with the formation of a new cabinet, to which three Radical Socialists were called. The bill raising the term of service to three years passed the Chamber, July 19, in spite of determined resistance.

On account of the extraordinary military expenses involved in the operations carried on in Morocco and in the installation of the additional class in the army, the government found it necessary to raise 1,300,000,000 francs ($247,000,000) by a loan; but they were defeated in the Chamber on a matter of detail in connection with this financial scheme, and resigned, December 2, 1913.

To appreciate fully the meaning of the Balkan conflicts for Russia, we must regard her diplomatic situation from a broader point of view. The agreement of Russia and Austria-Hungary on the eve of the Russo-Turkish War of 1877–8 relative to compensation for the Dual Monarchy has been interpreted as a virtual recognition by both parties of the existence of distinct Austrian and Russian spheres of interest in the Balkan peninsula, the former in the west, the latter in the east. But the action of the Teutonic powers in exercising influence eastward of the supposed line of division had long since made this agreement obsolete, if it had ever really existed. Each power, or group of powers, doubtless regarded the Balkan peninsula as an unrestricted field for diplomatic maneuvers, while the influence of events in raising or depressing the prestige of different powers was effective in all parts of the peninsula alike.

The vigorous prosecution of her policy in the Far East diverted Russia's attention in large measure from the Balkan peninsula for many years. But not long after the termination of the Russo-Japanese War, a change in the direction of Russian policy set in. On July 30, 1907, a convention establishing complete concord between Russia and Japan was signed at St. Petersburg. In commenting upon some speeches of Prince von Bülow about this time, the *Novoye Vremya*, an influential St. Petersburg paper, said that the "international ill-feeling towards Germany is explained, not by the envy of her neighbors, but by concrete facts........the unbroken record of German aggressiveness in Europe, Asia, and Africa. Under the mask of traditional friendship towards Russia, Teutonic statesmen incited the diplomatic novices, in whose hands the fate of our country was placed, to pursue a policy of enmity towards our Asiatic neighbors. The victor in the Russo-Japanese War proved once more to be Germany, who levied an indemnity on us in the form of a ruinous commercial treaty. It is this method of action that accounts for the moral isolation in which Germany finds herself."

Whatever measure of truth there is at the basis of this opinion, the fact of greatest significance is its acceptance and repetition by intelligent persons in Russia, creating an atmosphere of distrust, and a presumption of duplicity in considering Germany's conduct. Russia could now devote more attention to her very important interests in the Balkans. But the next year occurred Austria's high-handed annexation of Bosnia and Herzegovina without the previous authority of the signatory powers of the Treaty of Berlin. Russia's military and economic situation at the time probably did not permit her to adopt a determined attitude, and her prestige in the Balkans must have suffered a reverse in consequence. Russia could scarcely forget or forgive this humiliation. It was followed by a rapid development of her military and naval preparations. The influence of Russia was undoubtedly felt in the Balkans during the conflicts in 1912 and 1913. Russian diplomacy probably encouraged the formation of the league of the Balkan states. But, on the other hand, Russian diplomacy had receded before the danger of war, and permitted Serbia and Montenegro to be deprived of the most desirable fruits of their victories. The result was, that another diplomatic repulse in the Balkan peninsula would probably have destroyed Russia's prestige entirely and endangered those vital interests which we have already described.

Serbia had formally renounced her aspirations for a comprehensive national unity involving the incorporation of territories under the Hapsburg rule. But she could not so easily banish her disappointment. The victories in the Balkan Wars reanimated her hopes, and although Austria-Hungary again interposed limitations to her expansion, the dream of the Greater Serbia seemed to have been brought much nearer to realization, and undoubtedly patriotic societies, stimulated by hatred, carried on with increased ardor an annoying, and perhaps dangerous, agitation in the adjacent Austro-Hungarian territory. Let us consider a

little more closely the area upon which the Great Serbian structure was to be erected. If we run our eye westward from Roumania and Bulgaria upon the map, we notice in succession, Serbia, Bosnia, and Croatia and Dalmatia, all inhabited by people of similar stock, the Serbo-Croats. The Great Serbian aspirations would virtually be confined to these regions, to which, therefore, we may restrict our attention in endeavoring to estimate the danger involved for the Austro-Hungarian Monarchy.

The most striking events were taking place in Croatia. Croatia's political situation, with its subordinate diet or parliament, in relation to the Kingdom of Hungary might be compared with that of Ireland, as established by the Home Rule Bill, to the United Kingdom. The Croatian diet sends a delegation of forty representatives to the Hungarian Chamber of Deputies, and three to the Chamber of Magnates. The supreme importance of Croatia for Hungary consists in the fact that only through the dependent Croatian territory does Hungary enjoy uninterrupted communication with the sea.

We have remarked that Croatia escaped largely the process of Magyarization as applied extensively in the Hungarian dominions; but this is true only in a relative sense. For the spirit of opposition in Croatia developed chiefly out of complaints of attempted Magyarization. The governor of Croatia is called the Ban, and owes his appointment to the Hungarian Crown. The year 1903 was a turning point in Croat politics. Until then the government was usually successful in carrying out a policy prescribed in Budapest thanks to the limited suffrage and large proportion of official employés among the electors. Besides, the party which upheld nationalistic Croat aspirations did not coöperate with the Serbs. But in that year a coalition was organized, that is, one which represented the common sentiments and ideals of both the Croats and Serbs. This party very soon secured the passage of a law enlarging greatly the basis of suffrage, and in this way they assured for themselves an overwhelming majority in the diet. The government was powerless to effect anything against them by the regular constitutional means.

In March, 1908, the prosecution of fifty-three persons for high treason was commenced before a court in Agram, the capital of Croatia. They were charged with participation in a movement for separating Croatia, Slavonia, and Bosnia from Austria-Hungary and uniting them with Serbia. This trial, rendered famous by its political character, and the notably irregular character of the proceedings, continued until October 5, when thirty-one of the defendants were sentenced to various terms of penal servitude. They appealed against this verdict. It has been asserted that these proceedings were instigated by certain members of the government to make the contemplated annexation of Bosnia and Herzegovina appear more reasonable.

In December of the same year, Dr. Friedjung, a historian and friend of Count Aehrenthal, in an article in the *Neue Freie Presse of Vienna*, accused the Serbo-Croat coalition deputies in the Croatian diet of conspiring with Serbia. Thereupon the forty-nine deputies brought action for libel; and it was proved in the course of the trial that Dr. Friedjung's evidence, which had been supplied by the foreign office, was made up of forgeries.

Finally, in November, 1910, the appellate court reversed the treason verdict in Agram, and released the thirty-one prisoners who had been convicted in October, 1908. The higher court discovered that the court which had condemned them had made no effort to test the authenticity of the evidence.

Discouraged at the apparently hopeless majority of the

coalition in the election of January, 1912, the Ban then in office resigned, and M. de Tchuvaj was appointed to succeed him. He adopted a very drastic course, precipitating a constitutional crisis by dissolving the new diet before it had assembled, January 28. It appears that the newly-chosen deputies intended to pass a law of separation from the Kingdom of Hungary, and that one of them sent an address to the emperor, and to the heir presumptive, Archduke Francis Ferdinand, advocating a union of Croatia, Slavonia, Bosnia, Herzegovina, and Dalmatia to form a Slav kingdom under the House of Hapsburg. The members of the disbanded diet protested that the proclamation of dissolution was illegal, and the members of the coalition summoned a meeting of their partisans in Agram, February 6, for a demonstration in favor of renouncing the union with Hungary. The government had recourse to extreme measures. A royal rescript was published in the Official Gazette at Agram, April 3, conferring dictatorial authority upon Tchuvaj, who issued a proclamation the same day announcing the suspension of the Croatian constitution. At the same time the censorship of the press was established, and the right of assembly curtailed. These measures provoked intense indignation in Croatia, and societies were formed throughout the country to carry on the nationalist Croat propaganda. A student shot at Tchuvaj as he was passing in an automobile in Agram, June 8, but missed his aim, killing instead the director of education who was riding with the Ban.

The agitation in Croatia extended to Bosnia and Herzegovina, where Hungarian flags were burned as a demonstration of sympathy for the Croats. These provinces had received a considerable measure of autonomy, February 17, 1910, with a diet meeting at Sarajevo. The military commandant acts as head of the civil administration, under the ultimate authority of the Austro-Hungarian minister of finance. The diet passed a resolution of sympathy for the Croats, February 21. Constitutional government was not restored to Croatia until late in 1913.

It was commonly believed that the Archduke Francis Ferdinand favored the establishment of a Slav kingdom within the Hapsburg territories, involving the expansion of the Dual into a Triple Monarchy, by the admission of the new kingdom as a partner on terms of equality with Austria and Hungary. A prominent statesman has said that the Austro-Hungarian Monarchy is so necessary as an element in the European family of states, that if it did not exist, it would be created. But it is true, nevertheless, that a profound transformation is necessary before it can discharge effectively its function of usefulness. For, heretofore, its political energy has been largely dissipated in internal dissensions. This constant discord arises from the confinement of the national life of the various peoples within an unnatural political framework, from which they struggle to release themselves with greater determination as the consciousness of nationality grows stronger. There is no comprehensive movement tending to dissolve the general association of lands and peoples under Hapsburg rule. The aim of the most passionate struggles has usually been to secure liberation from such a relationship of political subordination as the subjection of the Croats to the Hungarians, the existence of which history alone can explain, but not justify. The substitution of a federal system of union would relax these galling and useless restrictions, and the adoption of the plan attributed to Archduke Francis Ferdinand would be the biggest step in the direction of federation. But the proposal of a federal union would doubtless encounter no less determined opposition at the present time than it did in 1861, for the predominant nationalities guard jealously their historic rights. Perhaps the only hope of salvation for Austria-Hungary by the introduction of a federal system has expired with the murdered Archduke Francis Ferdinand.

A fate as appalling and inexorable as that which appears in ancient Greek tragedy has pursued the House of Hapsburg in recent times. One by one the supports of the family have been stricken down, until the Emperor Francis Joseph is left at the age of eighty-four almost alone, at the twilight of life, in sorrow and gloom. His brother, Maximilian, was executed in Mexico, in 1867. His only son, Prince Rudolf, took his own life, in 1889. Eight years later, his sister-in-law perished in the burning of the Opéra Comique in Paris. His beautiful wife, the Empress Elizabeth, was assassinated in 1898 by an anarchist. His brother, Archduke John, has disappeared from public view.

The Archduke Francis Ferdinand was son of the Archduke Carl Louis, a brother of the Emperor Francis Joseph. He was born in 1864, and married, in 1900, a very beautiful and gifted Bohemian lady of Slavic race, the Countess Sophia Chotek, upon whom was subsequently conferred the title Duchess of Hohenberg. But Archduke Francis Ferdinand renounced for his children the right of succession to the throne because his wife was not of royal descent. Francis Ferdinand's vigorous, radical temperament had been the cause of anxious misgivings in the Dual Monarchy, where it was thought to portend a forceful policy of expansion. But a riper age had perhaps brought steadier counsels. On the other hand, his customary reserve of manner, passing at times into a state of melancholy or morbidness, and his association with the leaders of clericalism were regarded by some as evidences of a fanatical, or reactionary, disposition. The Slav policy, which has been ascribed to him, remains the most significant thing about him, and this must have been regarded with apprehension in precisely two quarters:

in Serbia, because a contented Slav kingdom in the Hapsburg realm would deprive the Greater Serbian hopes of their foundation, and in Hungary, because the fulfilment of the plan would necessarily deprive the Magyars of their Slav subjects.

The Archduke Francis Ferdinand left Vienna, June 23, 1914, to attend the military maneuvers in the province of Bosnia, as commander-in-chief of the Austro-Hungarian armies. On Sunday, June 28, accompanied by his wife, the Countess of Hohenberg, he paid a visit of ceremony to Sarajevo, the seat of the provincial administration. Sarajevo lies in a smiling, fertile plain, at the point where the Miljacka River emerges from a rugged, mountainous gorge. More than a third of the 45,000 inhabitants of Sarajevo are Mohammedans, and the association of various races in the same city, with their sharply contrasted forms of architecture, lends a picturesque charm to this meeting place of East and West. Sarajevo with its bright, prosperous appearance, its attractive streets, its oriental bazaar, and mosques with their slender minarets, did not deserve the ineffaceable stain of a horrible crime bringing in its train a catastrophe which will forever remain proverbial.

The archduke and duchess narrowly escaped being killed by a bomb thrown at their carriage, as they entered the town, by a youth, Nedeljko Cabrinovic, who injured thirteen bystanders in his murderous attempt. Later in the day they were shot by Gavrilo Prinzip with a Browning pistol, as they were riding back from a reception at the town hall. The crime was evidently the execution of a political conspiracy. It was assumed to be an act of revenge for the annexation of Bosnia and Herzegovina by Austria-Hungary, a step which Francis Ferdinand had heartily advocated. But it was also conceivable that the dominant motive of the conspirators was to prevent the fulfilment of the cherished projects popularly ascribed to the archduke. The complicity, or instigation, of one of the Serbian secret societies for political propaganda and terrorism was suspected at once.

For nearly a month the intentions of the Austro-Hungarian government were shrouded in impenetrable obscurity, although it was known that a court-martial at Sarajevo proceeded to the examination of Gavrilo Prinzip and twenty-one fellow conspirators. It was a period of ominous calm, of hushed excitement. All sorts of rumors were afloat. It was even whispered in Vienna that the awful murder had been instigated in Hungarian circles.

At length, like a thunderbolt from a clear sky, came the ultimatum of the Dual Monarchy to Serbia, July 23, demanding a satisfactory answer within forty-eight hours, which was practically an indictment of the whole Serbian nation for complicity before the fact, so far-reaching and peremptory in its character, as to reveal a deliberate intention of forcing war.

CHAPTER VI

Fruitless Attempts to Localize the Conflict

Austro-Hungarian note to Serbia. The extent of the Pan-Serbian danger to the Dual Monarchy. Motives of Austria-Hungary and attitude of Germany. The German communication to the powers defining her position; German view as to localizing the controversy. Russia's attitude. Count Berchtold denied that Austria-Hungary intended to annex Serbian territory. Sir E. Grey on the Austro-Serbian conflict. The request for an extension of the time limit for Serbia's reply. Sir E. Grey's idea of mediation by four powers. Reception of the Austro-Hungarian note in Serbia. Serbia's reply, July 25, and Austria-Hungary's objections to it. Rupture of diplomatic relations between Vienna and Belgrade. Sir E. Grey's proposal for an ambassadors' conference. Italy and France accepted it, Germany refused. Kaiser's return from Norway, July 26. Austro-Hungarian declaration of war against Serbia; and German imperial government's message to the German states, July 28. The Kaiser's exchange of telegrams with the Tsar. Russian partial mobilization. Russian formulas for an agreement with Austria-Hungary. Russian general mobilization. German ultimatum to Russia. German declaration of war against Russia, August 1. Germany demands statement of French attitude; declaration of war against France, August 3. The Kaiser's activity as mediator. Did Austria-Hungary adopt a conciliatory attitude at the last moment?

Austria-Hungary regarded the consequences of the second Balkan War and the resulting elation of the Serbs as a constant menace. She had undoubtedly been observing the conduct of Serbia very closely since the Treaty of Bucharest; and any reckless or imprudent step on the part of the Slav neighbor, which could in any way be interpreted as a provocation, was certain to lead to Austrian intervention. Assuming that active measures to repress the spirit of the Serbs were desired by a large and influential element in Austria-Hungary —and of the truth of this conjecture there appears to be little doubt—we may conclude that the deplorable crime of Sarajevo played directly into the hands of this more energetic party. Public sentiment generally was favorable to Austria-

Pashitch, Serbian Prime Minister.

Peter, King of Serbia.

Alexander, Crown Prince of Serbia.

Hungary. The affliction of the aged emperor excited universal sympathy. The powers were prepared to uphold Austria-Hungary in any reasonable steps which she might consider it necessary to take for protecting herself against similar outrages. It was to be assumed that the sentiment of monarchical solidarity would assure the unhesitating support of William II of Germany, who had been unwilling to support Austria-Hungary in extreme measures the year before.

The Austro-Hungarian foreign office studied the situation deliberately. The view was expressed up to the last, by those who shared in its counsels, that there was no reason to anticipate a serious turn in affairs. Apprehension was quieted to such a degree that the British and Russian ambassadors were absent from Vienna and Berlin at the time when the Austro-Hungarian note was finally presented to Serbia.

In the mean time the findings of the criminal investigation in Sarajevo pointed to the powerful Pan-Serbian society Narodna Odbrana as the instigator of the crime, and implicated persons in high places in Serbia. Serbia's record was such as to afford basis for a presumption in support of the charges which Austria-Hungary brought against her. Her annals contain many regicides. The most recent was the assassination of King Alexander, whose policy had been favorable to Austrian interests, together with his queen, Draga, by a party of officers, in 1903, a crime that was hideous in its details. Thus the present King of Serbia, Peter Karageorgevitch, owes his throne to political murder.

On the other hand, we must not lose sight of the fact that the assassin of Sarajevo and his two chief fellow-conspirators, all youths under twenty years of age, were Austro-Hungarian subjects. Another circumstance of importance in forming a judgment of Serbia's position: six days before the crime was committed, the Serbian minister in Vienna warned the Austrian government that the archduke's visit to Bosnia would be perilous on account of the probable existence of a plot against his life. Austria-Hungary demanded that Serbia accept the findings of the Sarajevo investigation, and in consequence adopt proposals which seemed to threaten her independence, without an opportunity of examining the evidence. But was the record of Austro-Hungarian jurisdiction in cases involving political questions a guarantee for absolute confidence in the soundness of the results of the Sarajevo investigation? It could scarcely be maintained that it was such a guarantee. A consideration of the character of the Agram and Friedjung trials tends to justify Serbia's hesitation in accepting without independent enquiry the conclusions reached by the Sarajevo court.

The note began by reminding the Serbian government of the declaration made through its minister in Vienna, March 31, 1909, that Serbia would conform to the decision of the powers relative to Bosnia, cease her attitude of protest and resistance, change the direction of her policy towards Austria-Hungary, and live in future on good neighborly terms with the latter. The Austro-Hungarian government affirmed that instead of fulfilling these promises Serbia had done nothing to suppress a movement aiming to detach certain territories from Austria-Hungary, which found expression beyond the territory of the Kingdom of Serbia in a series of outrages culminating in the horrible crime at Sarajevo, June 28. The participation of Serbian officers and officials in the intrigues of the societies which directed their agitation against the Austro-Hungarian Monarchy, the violence of the Serbian press, and the unwholesome propaganda in education were proof of the culpable tolerance of the Serbian government. The evidence and confessions of the perpetrators of the outrage of June 28 indicated that the murder was planned in Belgrade, that the murderers received their arms and explosives from Serbian officers and officials, and that their passage across the boundary was facilitated by the Serbian frontier service. The Austro-Hungarian government felt bound to put an end to these menacing intrigues, and for this purpose they demanded that the following denunciation of the subversive agitation be published by the Serbian government in its *Official Journal* for July 26:

"The Royal Government of Serbia condemns the propaganda directed against Austria-Hungary, that is to say, the general tendency of which the final aim is to detach from the Austro-Hungarian Monarchy territories belonging to it, and they sincerely deplore the fatal consequences of these criminal proceedings.

"The Royal Government regrets that Serbian officers and functionaries have participated in the above-mentioned propaganda and thus compromised the good neighborly relations to which the Royal Government was solemnly pledged by its declaration of March 31, 1909.

"The Royal Government, which disapproves and repudiates all idea of interfering or attempting to interfere with the destinies of any part whatsoever of Austria-Hungary, considers it its duty formally to warn officers and functionaries, and the whole population of the kingdom, that henceforward it will proceed with the utmost rigor against persons who may be guilty of such machinations, which it shall use all its efforts to anticipate and suppress."

"This declaration shall simultaneously be communicated to the Royal Army as an order of the day by His Majesty the King and shall be published in the Official Bulletin of the army.

"The Royal Serbian Government further undertakes:

1. To suppress any publication which fosters hatred of,

and contempt for, the Austro-Hungarian Monarchy, and whose general tendency is directed against the latter's territorial integrity.

2. "To proceed at once with the dissolution of the society Narodna Odbrana, to confiscate its entire means of propaganda, and to proceed in the same manner against the other societies and associations in Serbia which occupy themselves with the propaganda against Austria-Hungary. The Royal Government will take the necessary measures so that the dissolved societies may not continue their activities under another name or in another form.

3. "To eliminate without delay from the public instruction in Serbia, both as regards the corps of instructors and the means of instruction, that which serves, or may serve, to foster the propaganda against Austria-Hungary.

4. "To remove from military service and the administration in general all officers and officials who are guilty of propaganda against Austria-Hungary, and whose names, with a communication of the material which the Imperial and Royal Government possesses against them, the Imperial and Royal Government reserves the right to communicate to the Royal (Serbian) Government.

5. "To consent that in Serbia officials of the Imperial and Royal Government coöperate in the suppression of a movement directed against the territorial integrity of the Monarchy.

6. "To commence a judicial investigation against the participants of the conspiracy of June 28th, who are on Serbian territory. Officials delegated by the Imperial and Royal Government will participate in the examinations.

7. "To proceed at once to arrest Major Voija Tankosic and a certain Milan Ciganovic, a Servian state employé, who have been compromised by the results of the investigation at Sarajevo.

The Tsar of Russia and William II on the way to a hunt in Germany.

8. "To prevent through effective measures the participation of the Serbian authorities in the smuggling of arms and explosives across the frontier, and to dismiss those officials of Shabatz and Loznica, who assisted the perpetrators of the crime of Sarajevo in crossing the frontier.

9. "To give to the Imperial and Royal Government explanations in regard to the unjustifiable utterances of high Serbian functionaries in Serbia and abroad who have not hesitated, in spite of their official position, to express themselves in interviews in a hostile manner against Austria-Hungary after the outrage of June 28th.

10. "The Imperial and Royal Government expects a reply from the Royal Government at the latest by 6 o'clock on Saturday evening, July 25th. A memorandum dealing with the results of the investigation at Sarajevo, so far as they concern points 7 and 8, is enclosed with this note."

The Austro-Hungarian diplomatic representatives in the other countries were requested to communicate the contents of this note to the powers to which they were individually accredited, accompanying them with the following observations as further explanation of Austria's action: that directly after March 31, 1909, Serbia became a center for criminal agitation, that the societies for creating disorders in Austria-Hungary included persons in the highest ranks of official and unofficial society, that the press made itself a serviceable organ for the propaganda, and that since the recent Balkan crisis members of bands formerly operating in Macedonia had placed their services at the disposal of the propaganda against Austria-Hungary, and that the Serbian government had not taken the slightest step to put a stop to all this lawless conduct.

"The patience of the Imperial and Royal Government in the face of the provocative attitude of Serbia was inspired by the territorial disinterestedness of the Austro-Hungarian

Monarchy and the hope that the Serbian government would end in spite of everything by appreciating Austria-Hungary's friendship at its true value. By observing a benevolent attitude towards the political interests of Serbia, the Imperial and Royal Government hoped that the Kingdom would finally decide to follow an analogous line of conduct on its own side. In particular, Austria-Hungary expected a development of this kind in the political ideas of Serbia, when, after the events of 1912, the Imperial and Royal Government by its disinterested and ungrudging attitude, made such a considerable aggrandizement of Serbia possible."

But this benevolent attitude had exercised no restraining effect on the conduct of the Kingdom of Serbia, which continued to tolerate the propaganda on its territory with the fatal consequences of June 28th. Then the Austro-Hungarian government had been compelled to take the present urgent step for its own safety.

An annex attached to the communication stated the conclusions of the criminal enquiry of the court of Sarajevo, which were briefly as follows:

The plot was formed in Belgrade by Gavrilo Prinzip, Nedeljko Cabrinovic, Milan Ciganovic, and Trifko Grabez, with the assistance of Major Voija Tankosic.

The six bombs from the arms depôt of the Servian army at Kragujevac and four Browning pistols, with which the crime was committed, were delivered by Ciganovic and Tankosic to Prinzip, Cabrinovic, and Grabez at Belgrade.

By a secret system of transport organized by Ciganovic, and with the complicity of the frontier officers at Shabatz and Loznica, Prinzip, Cabrinovic, and Grabez had been able to cross the frontier of Bosnia and smuggle in their contraband of arms.

Professor Hans Delbrück, who occupies a famous chair of history in the University of Berlin, acknowledges that

Austria-Hungary demanded conditions which would have placed Serbia under her permanent control, and he justifies this action on the ground that the Dual Monarchy was moved by the instinct of self-preservation. Our judgment of the conduct of the Austro-Hungarian government may involve, therefore, a consideration of the extent of the peril to which the monarchy was exposed by reason of the Greater Serbian propaganda. The Serbian societies for agitation probably aimed at the detachment of Croatia and Dalmatia as well as Bosnia and Herzegovina from the Hapsburg realm, and their incorporation with Serbia to form the Greater Serbia which was their cherished dream. The consequent danger is magnified, if we assume that this Greater Serbia would be virtually an outlying part of a Pan-Slavist Russia. It is doubtful, however, whether a compact state of nearly 10,000,000 Serbo-Croats would be any more subservient to Russia than the 2,000,000 Bulgarians proved themselves to be in the 80's of the last century. But on the other hand, we must associate in our minds the danger arising from the Greater Serbian aspirations with the menace of Italian Irredentism. For the fulfilment of both these programs would completely absorb the Austro-Hungarian seaboard, excluding the Dual Monarchy from access to the Adriatic. In fact, in their utmost territorial extension the aspirations of the two nationalities may be said to overlap. It would, of course, be intolerable for a realm like the Austro-Hungarian Monarchy to be deprived entirely of its coastline, and thus be reduced, in this respect, to the present situation of Serbia. That would be destructive of the vital interests of the monarchy.

But the question arises, whether the realization of the Greater Serbian ideal with such an extension is possible. The imagined function of the present Kingdom of Serbia as nucleus for the proposed formation of a very much more extensive national state has been compared with that of Piedmont at the beginning of the movement for Italian unity. This comparison has doubtless encouraged the ambitious Serbs. Both movements threatened the integrity of the Austrian Empire. But Professor Delbrück points out that circumstances are not favorable to a repetition by Serbia of the rôle of Piedmont as leader in a comprehensive movement for national unification. For the community of national ideals necessary for the creation of a national state does not exist among the Serbo-Croats. The two branches of this stock hate each other, according to the eminent Berlin authority, and the "majority of the Southern Slavs do not want to hear anything about Greater Serbia," and "unlike the former Lombardo-Venetians, they are true and loyal subjects of the House of Hapsburg." This is doubtless true; for we have already observed that the aim of Croatian agitation was to set up a new Slavic kingdom under the rule of the House of Hapsburg with the intention that it should enter as a third member into the political partnership of the present Dual Monarchy. Professor Delbrück might have added as further evidence for the lack of compatibility of the Serbo-Croats the fact that the Serbs are mostly adherents of the Eastern Orthodox Church and employ the same alphabet as the Russians, while the Croats and Dalmatians are Roman Catholics and use the Latin alphabet. These two elements of dissimilarity alone would be sufficient to render a harmonious national union almost impossible. Such facts limit the practical extent of the Greater Serbian agitation, and the consequent danger to the Austro-Hungarian Monarchy, to Bosnia and Herzegovina, where the Serbs constitute the greater part of the population.

It follows, therefore, that Austria-Hungary really took over this problem with the provinces Bosnia and Herzegovina. The Greater Serbian movement would probably

House of Parliament, Vienna.

Houses of Parliament, Budapest.

have resulted, at most, in the detachment of these provinces from the Dual Monarchy. This would doubtless have been a humiliation, as such losses of territory are commonly regarded even to-day in the lives of nations. The real character of such an occurrence must be judged by the inclination of the inhabitants. For history offers abundant evidence to prove that the retention of unwilling populations is not a source of strength. We conclude that the Greater Serbian movement would not have imperilled the very existence of the monarchy, as has been so often asserted.

Although the note of the Austro-Hungarian government was transmitted to the Serbian government at six o'clock, Thursday evening, July 23, it was not until the next day that Austria's action, the text of the note, and the explanatory remarks, which have already been mentioned, were communicated to the powers. This left a very short time for international consultation or for imparting advice to Serbia, and it is very likely that this was precisely the purpose of the Austro-Hungarian government in making the time limit so brief. They doubtless hoped that the incisive, peremptory terms would have the effect of bringing forth an immediate, and practically complete, rejection of the note or else submission to their demands, excluding in this way further negotiations and uncertainty. Austria-Hungary was determined to make an end once for all of the Serbian propaganda and of Serbian opposition to her policy. The Austro-Hungarian authorities had probably determined that at least an armed demonstration on Serbian territory was necessary for the maintenance of their prestige. Popular feeling in Austria as well as the calculations of diplomacy seem to have anticipated a rejection of the demands of the note by Serbia. The British Ambassador in Vienna, in one of his despatches says of the note: "Its integral acceptance by Serbia was neither expected nor desired, and when, on the following afternoon, it was at first rumored in Vienna that it had been unconditionally accepted, there was a moment of keen disappointment."

It is scarcely conceivable that Austria-Hungary took this momentous action, which would almost certainly involve her in hostilities with Serbia, and would at least lead to danger of a general European war, without a complete previous understanding with Germany as to the nature and means of her policy, and special assurance of German solidarity. Whether the actual text of the note had been communicated to the Foreign Office at Berlin before it was despatched to Belgrade is less certain. The German government formally disclaimed any previous knowledge of the contents of the note; but this statement, if true, is probably true in only a strictly literal sense. The German government lost no time, directly the note had been transmitted, in giving it their hearty endorsement, and in announcing firmly and unequivocally their attitude with respect to the Austro-Serbian quarrel. Even before the note had been sent, on July 22, Sir H. Rumbold, British *chargé d'affaires* at Berlin, telegraphed Sir Edward Grey that Herr von Jagow, German Secretary of State for Foreign Affairs, in alluding to the action which Austria was about to take, had "insisted that the question at issue was one for settlement between Serbia and Austria alone, and that there should be no interference from outside in the discussion between these two countries. He had therefore considered it inadvisable that the Austro-Hungarian government should be approached by the German government on the matter........"

Almost simultaneously with the transmission of the Austro-Hungarian note, or ultimatum, to Serbia, the German Imperial Chancellor sent instructions to the German ambassadors at the capitals of the *entente* powers, indicating very clearly the attitude of the German government, and instructing them to communicate it to the governments to which each was accredited. This message declared that the publications of the Austro-Hungarian government concerning the assassination of the Austrian successor to the throne and his consort proved that Belgrade was the center of a propaganda for separating the southern Slavic provinces from the Austro-Hungarian Monarchy; that Pan-Serb chauvinism all but precipitated war in 1909; that instead of observing the promise of good behavior made at that time, the Serbian government had tolerated the agitation, which led to the existing crime and permanently threatened the safety of Austria-Hungary; and that for these reasons the demands of the monarchy were justifiable, and it must renounce its position as a great power unless it were prepared to enforce them by every means. The communication continued:

"I have the honor to request you to express yourself in the sense indicated above to (the present representative of M. Viviani) (Sir Edward Grey) (M. Sazonoff) and therewith give special emphasis to the view that in this question there is concerned an affair which should be settled solely between Austria-Hungary and Serbia, the limitation to which it must be the earnest endeavor of the powers to insure. We anxiously desire the localization of the conflict because every intercession of another power on account of the various treaty-alliances would precipitate inconceivable consequences.

"I shall look forward with interest to a telegraphic report about the course of your interview."

It may be observed that this communication was sent off by the German Imperial Chancellor on the 23rd, although the Austro-Hungarian government did not inform the other powers of their action at Belgrade until the next day. It was evidently the intention of the German government that their very positive representations, backing up the Austrian position, should reach the foreign offices of the *entente* powers at just about the same time as the information from Vienna regarding the action of the Austro-Hungarian government.

The German attitude was thus very clearly stated, and there is no evidence to prove that it was in any way altered during the whole period of negotiation and discussion. Germany insisted that the question at issue was one for settlement between Austria-Hungary and Serbia alone, and that there should be no interference from outside in the discussions between these two countries. Germany assumed that Austria-Hungary's promise not to annex any Serbian territory was sufficient to put at rest any reasonable motives for uneasiness on the part of other powers. It may be observed, however, that by crushing Serbia's military power of resistance, or imposing an overwhelming war indemnity upon her, or by some other form of interference, Austria-Hungary might have reduced Serbia to the position of a vassal, while all the time ostensibly respecting her integrity with the utmost scrupulousness. History offers abundant examples of such proceedings. The representations of Germany probably aroused, rather than diminished, the apprehension of Russia. Once Serbia should begin to move as a satellite in the orbit of Austria-Hungary, Russian prestige in the Balkans would experience a sudden and fatal decline; and it is not inappropriate to mention in this connection, that Count Mensdorff, the Austrian Ambassador in London, remarked to Sir Edward Grey that Serbia had heretofore been regarded as within the Austrian sphere of influence. Germany's frequently repeated desire for a localization of the quarrel in the interests of peace signified nothing more or less than the demand that Austria-Hungary should be left to deal with Serbia without interference from other powers.

Immediately upon receiving information of the Austrian action in Belgrade and the German communication relative to it, M. Sazonoff, the Russian Foreign Minister, declared that Russia could under no circumstances permit the matter to be settled alone between the parties directly concerned. We need only recall how popular enthusiasm in Russia was the decisive factor in causing the Russo-Turkish War in 1877, and then the nature of the occurrences in the Balkans in 1909, and 1912–1913, to appreciate the fact that public opinion would undoubtedly be an important factor in Russia during this crisis, when it might appear that the independence of fellow Slavs was being threatened. The attitude of Russia from the first, therefore, was opposed to that of Germany. Still Russia's action was unquestionably moderate, at least at the beginning. M. Sazonoff immediately telegraphed to the Russian *chargé d'affaires* at Vienna instructing him to request that the period allowed for the Serbian reply should be extended, to give the powers time to study the results of the judicial enquiry at Sarajevo, so that they could intelligently offer their advice to the Serbian government. Similar communications were likewise despatched to Germany, Great Britain, France, and Italy, begging them to urge an extension of the time limit. Now it will appear at each successive stage of the discussions that the conflict between the attitude of Germany and Austria and that of Russia regarding the extent of the interests involved in the Austro-Serbian difficulty is fundamental for all the issues arising in the exchange of views by powers. Thus, for example, if the Austro-Serbian quarrel concerned only the two original parties, Austria was under no obligation to give the other powers time to consult; but if it was a question involving general European interests, then the very brief time limit was distinctly a slight to the other great powers, since it tended to exclude them from participation.

Count Berchtold explained the Austro-Hungarian attitude to the Russian *chargé d'affaires* on the 24th, before the latter had received any instructions from St. Petersburg. He emphasized particularly the assertion that Austria-Hungary would lay claim to no Serbian territory, and he declared that he had no intention of bringing about a change in the balance of power in the Balkans. Austria-Hungary's step was intended as a definite means for checking the Serbian intrigues. Count Berchtold departed for Ischl, the summer residence of Emperor Francis Joseph, early on the 25th, and his absence from Vienna at this critical period seems very noteworthy, at least. On July 25th, the Russian *chargé d'affaires* at Berlin reported that Herr von Jagow, the Foreign Secretary, was afraid that in the absence of Count Berchtold the Russian request for an extension of time would produce no results; and the same day, the Russian *chargé d'affaires* at Vienna telegraphed to Ischl, receiving in reply an official refusal of the Russian request.

Directly after receiving the Austrian and German communications on the 24th, M. Sazonoff had conferred with the French and British ambassadors at St. Petersburg, expressing the view that Russia could not be indifferent to Austria's action in humiliating Serbia. The French ambassador promised that France would perform her obligations as an ally of Russia, and both he and M. Sazonoff urged upon Sir G. Buchanan the expediency of a British declaration of solidarity with France and Russia in this crisis, declaring that if Great Britain would be firm, there would be no war. On the following day, M. Sazonoff told Sir G. Buchanan that if Serbia should appeal to the powers, Russia would be willing to stand aside and leave the question in the hands of Great Britain, France, Germany, and Italy.

When Sir Edward Grey was informed by the Austrian ambassador of the contents of the note to Serbia, he exclaimed that he had never "seen one state address to another independent state a document of so formidable a character." He thought the acceptance of the fifth demand would hardly be consistent with Serbia's independent sovereignty. Sir Edward Grey immediately telegraphed to Sir H. Rumbold, British *chargé d'affaires* at Berlin, stating that if the Austrian note to Serbia did not lead to trouble between Austria and Russia, he had no concern with it, but he was apprehensive of the view Russia would take. He went on to say that the only chance of a mediating influence being effective would consist in the four powers, Germany, Italy, France, and Great Britain working together at Vienna and St. Petersburg in favor of moderation. Sir H. Rumbold was instructed to communicate this to the German Foreign Secretary. Likewise Sir Edward Grey instructed the British *chargé d'affaires* in Vienna to support the Russian request for an extension of the time limit for Serbia's reply to the note. Herr von Jagow, German Secretary of State for Foreign Affairs, promised on July 25, that if relations between Austria-Hungary and Russia became more acute, he was ready to fall in with Sir E. Grey's suggestion that the four powers work together in favor of moderation at Vienna and St. Petersburg. At the same time Russia expressed her willingness to stand aside and submit the controversy to the four powers mentioned. Sir Edward Grey worked from first to last with the idea in mind of a mediation, or conference, of the four powers not directly concerned in the original conflict. This policy was suggested by the experience of the ambassadors' conferences in London during the Balkan crisis the year before, which had undoubtedly contributed to the maintenance of the general peace.

Prime Minister Pashitch was not in Belgrade at the time when the Austro-Hungarian note was delivered, July 23. He did not return to the capital until the next day, and in the mean time the minister of finance begged the Russian *chargé d'affaires* for assistance. On the 24th, the prince regent of Serbia addressed the following telegram to the Tsar:

"Yesterday evening the Austro-Hungarian government delivered a note to the Serbian government concerning the conspiracy at Sarajevo. Appreciating her international obligations, Serbia had declared from the first that she condemned the horrible crime and that she was ready to institute judicial proceedings within her own territory, if the complicity of certain of her own subjects should be proved in the course of the process conducted by the Austro-Hungarian authorities. Nevertheless, the demands contained in the Austro-Hungarian note are needlessly humiliating for Serbia and incompatible with her dignity as an independent state. For instance, they demand in a peremptory manner a declaration by our government in the Official Bulletin, and an order of the sovereign to the army, expressing our intention of repressing the spirit of hostility against Austria and alluding to our criminal weakness with regard to our perfidious actions. They demand, likewise, the introduction of Austro-Hungarian functionaries into Serbia to coöperate with our own officials in the investigation, and to superintend the execution of the other conditions indicated in the note. We are given a space of forty-eight hours in which to accept all, and failing this, the Austro-Hungarian legation will leave Belgrade. We are ready to accept such of the Austro-Hungarian conditions as are compatible with our position as an independent state, together with such as Your Majesty shall advise; all persons will be severely punished whose participation in the conspiracy will be proved. Some of the demands cannot be executed without changes in our legislation, a proceeding which requires more time. The time granted us is too brief. It is possible

Belgrade, capital of Serbia, from the Hungarian bank of the Danube.

The Royal Palace at Belgrade.

that we shall be attacked at the expiration of the time limit by the Austro-Hungarian army which is now being concentrated on our frontier. We cannot defend ourselves and we beg Your Majesty to grant us your assistance as soon as possible..........."

On the afternoon of Saturday, July 25, Serbia returned to Austria-Hungary a reply, which appeared to all, except the Teutonic powers, to be practically an entire acceptance of the demands which had been submitted, subject to delays necessary for changes in the Serbian constitution and legislation, and subject to an explanation by the Austro-Hungarian government of the precise intention of the proposal for the participation of officials of the Dual Monarchy in the investigations in Serbia. It might seem, at least, to have gone far enough to form a basis for further negotiation. Sir Edward Grey received a draft of it a short time before it was transmitted to the Austro-Hungarian government, and immediately expressed the hope to Germany that she would urge the Austro-Hungarian government to accept it; but the Berlin government seems merely to have passed on the expression of Sir Edward Grey's hope to the Austro-Hungarian government through their ambassador in Vienna without added pressure.

Nothwithstanding the apparently conciliatory character of the Serbian reply to the Austro-Hungarian note, the Vienna government promptly declared that it was unsatisfactory, simply a play for time. One cannot be surprised that the Serbian government, when summoned to reply within forty-eight hours to a series of demands which Austria-Hungary might have been carefully weighing for the greater part of a month, may have displayed a tendency to gain time for a fuller consideration of these unexpected requirements by avoiding too specific acceptance of all the conditions.

The following is an abbreviated version of the Serbian reply, each part accompanied by a summary of the Austro-Hungarian objections:

Serbia directed attention to the fact that the expressions of protest by representatives and members of her government had ceased since she formally renounced her attitude of protest in her note to the Austro-Hungarian government, March 31, 1909, and that no attempt had been made since that time by the Serbian government or its organs to change the existing state of political affairs in Bosnia and Herzegovina. The Serbian reply attested, moreover, that the Austrian government had never made any representation on the basis of Serbia's agreement, except one concerning a school book, and on that occasion the Austro-Hungarian government had received an entirely satisfactory explanation. It was thanks to Serbia's moderation in the Balkan crisis that European peace had been preserved. Serbia was not responsible for manifestations of a private character, as newspaper articles and the activity of societies. The Serbian government was "painfully surprised by the assertions that subjects of Serbia had participated in the preparations of the outrage in Sarajevo. The government had expected to be invited to coöperate in the investigation of all that concerns this crime, and it was ready, in order to prove its entirely correct attitude, to proceed against all persons in regard to whom it should receive information." The Serbian government was ready to hand over for trial any Serbian subject of any rank "for whose complicity in the crime of Sarajevo it should have received proof."

The Austro-Hungarian government objected, in connection with this portion of Serbia's note of reply, that the allusion to her official faultlessness was calculated to obscure the real issue, inasmuch as Austria-Hungary had not remonstrated against positive transgressions but against sins of omission, that is, the failure to suppress the agitation against the integrity of the Dual Monarchy. According to the Austro-Hungarian observations on this part of the note, not only do the institutions of modern states commonly subject the press and societies to state control, but there was provision for the exercise of this control in Serbia's legislation. Furthermore, the Serbian government had been informed that suspicion rested on certain individuals within its territory, and should have instituted an investigation spontaneously.

The Serbian government pledged itself to cause to have published in the *Official Journal*, on July 26, a declaration which repeated the text dictated by the Austro-Hungarian government with some minor alterations, namely, *any* substituted for *the* (*toute* for *la* in the original French text), and two interpolations indicated in the following extract by ( ): "The Royal Government of Serbia condemns (the *Austrian text*), any *Serbian reply*) propaganda (which may be) directed against Austria-Hungary, that is to say, the general tendency of which the final aim is to detach from the Austro-Hungarian Monarchy territories belonging to it, and they sincerely deplore the fatal consequences of these criminal proceedings. The Royal Government regrets that Serbian officers and functionaries have participated (according to the communication from the Imperial and Royal Government) in the above mentioned propaganda, and thus compromised the good neighborly relations to which the Royal Government were solemnly pledged by their declaration of March 31, 1909."

The remainder is identical with the text prescribed by the Austro-Hungarian government, as already given on page 207 except that the words, "The Royal Government," have been unintentionally omitted, as it would appear, from the beginning of the next sentence, a silent testimony, perhaps, to the conditions of anxious haste under which the Serbian reply was prepared.

Austria-Hungary objected, with reference to Serbia's promised official declaration, that the first change was intended to imply that a propaganda did not exist, and was therefore proof of insincerity, as affording the subterfuge for later occasions that Serbia had not specifically acknowledged the existence of the existing propaganda and disavowed it. Austria-Hungary likewise expressed her dissatisfaction at the insertion of the words "according to the communication from the Imperial and Royal Government" on the same ground, claiming that they were added merely as a trick, to preserve for Serbia a free hand in future.

1. "To introduce at the first regular convocation of the Skupschtina a provision into the press law providing for the most severe punishment of incitement to hatred or contempt of the Austro-Hungarian Monarchy, and for taking action against any publication the general tendency of which is directed against the territorial integrity of Austria-Hungary. The Government pledges itself at the approaching revision of the constitution to cause an amendment to be introduced into article 22 of the constitution of such a nature that such publications may be confiscated, a proceeding at present impossible under the categorical terms of article 22 of the constitution."

Austria-Hungary declared that these proposals were entirely unsatisfactory and evasive. She had not demanded the passage of laws which might never be enforced, but Serbia's pledge effectively to suppress intriguing publications. Serbia promised to pass a press law permitting the government to punish expressions hostile to Austria-Hungary, and an amendment to the constitution to authorize the government to confiscate such publications; but she offered no guarantee that such legislation would be strin-

Sir Maurice William Ernest de Bunsen, British Ambassador

Count Albert von Mensdorff-Pouilly, Austro-Hungarian

gently enforced, when once it had been duly enacted.

One naturally wonders, in reading Austria-Hungary's objection, whether a literal acceptance of the first demand, that is, a solemn promise of the Serbian government to act in violation of its own constitution, would have afforded a more reliable guarantee of future conduct. Any guarantee short of intervention must depend on the promise of the Serbian government, and, if the Austro-Hungarian government had no faith in the pledges of the Serbian government, was not its own conduct rather disingenuous in demanding promises which would not in any case be satisfactory?

2. "The Government possesses no proof, nor does the note of the Imperial and Royal Government furnish them with any, that the Narodna Odbrana and other similar societies have committed up to the present any criminal act of this nature through the proceedings of any of their members. Nevertheless, the Royal Government will accept the demand of the Imperial and Royal Government, and will dissolve the Narodna Odbrana Society, and any other society which may be directing its efforts against Austria-Hungary."

Austria-Hungary objected that this was only a half-way concession, because they had demanded that Serbia should confiscate the means of propangada of these societies, and prevent them from reorganizing under other names and forms. Serbia's reply was really, therefore, no complete guarantee for putting an end to the agitation.

3. "The Royal Serbian Government undertakes to remove without delay from public educational establishments in Serbia all that serves or could serve to foment propaganda against Austria-Hungary, whenever the Imperial and Royal Government will furnish it with the facts and proofs of this propaganda."

Austria-Hungary remarked that Serbia first demanded proofs of a hostile propaganda in the public instruction, although it must know that the text-books contained objectionable material, and that a large portion of the teachers belonged to the societies for agitation. The Serbian government did not mention the elimination of the hostile element from the teaching force.

4. "The Royal Government also agrees to remove from military service all such persons as the judicial enquiry will have proved to be guilty of acts directed against the integrity of the territory of the Austro-Hungarian Monarchy, and they expect the Imperial and Royal Government to communicate to them at a later date the names and the acts of these officers and officials for the purposes of the proceedings which are to be taken against them."

Austria-Hungary observed that the Serbian government made its promise to dismiss officers and functionaries conditional upon a judicial investigation for establishing their guilt in actions against the territorial integrity of the Austro-Hungarian Monarchy. She asserted, furthermore, that this limitation was equivalent to the non-fulfilment of her demand, because such officers and functionaries could not be condemned by a judicial procedure, inasmuch as propaganda is not punishable by law in Serbia. The reply was, therefore, elusive.

But it seems more likely that the introduction by Serbia of this limitation in the scope of the concession as required by Austria-Hungary was due to a much more cogent motive than the intention of eluding the purpose of the demand by a mere evasive trick. Austria-Hungary had reserved to herself the right to submit the names of those officers and functionaries who should be dismissed for complicity in the agitation. But this would have made the whole military and civil service of Serbia dependent upon the discretion of Austria-Hungary, who would have been in a position to paralyze the action of the entire administration by causing the elimination of all the most able officers and functionaries on the alleged grounds of complicity in the propaganda. The reservation by Serbia of the privilege of an investigation previous to their dismissal was an indispensable safeguard for her national independence.

5. "The Royal Government must confess that it does not clearly grasp the meaning or the scope of the demand made by the Imperial and Royal Government that Serbia shall undertake to accept the collaboration of the organs of the Imperial and Royal Government upon their territory, but they declare that they will admit such collaboration as agrees with the principles of international law, with criminal procedure, and with good neighborly relations."

Austria-Hungary objected that this apparent compliance was misleading. "The international law, as well as the criminal law, has nothing to do with the question; it is purely a matter of the nature of state police which is to be solved by way of a special agreement. The reserved attitude of Serbia is therefore incomprehensible, and on account of its vague general form it would lead to unabridgeable difficulties."

But it must be observed that the nature of the demand itself was vague and general, very likely intentionally so, and capable of a very far-reaching application. It might conveniently have served as an entering wedge for a permanent Austrian administrative penetration into the affairs of Serbia. Here, less than elsewhere, can we be surprised at an attitude of reserve on Serbia's part.

6. "It goes without saying that the Royal Government consider it their duty to open an enquiry against all such persons as are, or eventually might have been, implicated in the plot of the 28th of June, and who happen to be within the territory of the kingdom. As to the participation in this enquiry of Austro-Hungarian agents or authorities

appointed for this purpose by the Imperial and Royal Government, the Royal Government cannot accept such an arrangement, as it would be a violation of the constitution and of the law of criminal procedure; nevertheless, in concrete cases communications as to the results of the investigation in question might be given to the Austro-Hungarian agents."

The Austro-Hungarian government asserted that in this part of the reply the Serbian government was guilty of a deliberate misconstruction of the demand. Austria-Hungary had demanded that Serbia institute judicial proceedings (*enquête judiciaire*) at once against the accessories of the plot of June 28, and that Austro-Hungarian delegates should collaborate in the investigations (*recherches*) relating thereto; in other words, Austro-Hungarian delegates would share in the police investigations, but not in the judicial proceedings. By the use of the term *enquête* in both instances it might appear that Serbia understood the sense of the demand to be that Austro-Hungarian delegates should intervene in the judicial proceedings. Therefore, Austria-Hungary declared that Serbia was trying to justify her refusal by making the demands seem impossible.

The distinction of terms upon which this argument is based has been regarded by some writers as unreasonably petty. There seems, however, that there is a real difference between the two expressions. The replies of Serbia to the fifth and sixth demands are possibly, but not necessarily, elusive or evasive in their nature. These two demands, together with the fourth, constitute, undoubtedly, the most weighty part of the Austro-Hungarian note. The fundamental problem in connection with the whole note is suggested by the discussion of these three headings. Was the Austro-Hungarian government justified in making such demands of an independent state, even though it may have had very serious provocation?

The Austro-Hungarian government claimed, however, that precedents were not wanting for precisely the kind of police collaboration which it had really demanded. According to their view, the feigned misinterpretation and refusal of Serbia had as motive, not so much general principles, as simply the apprehension of embarrassing results from such an international investigation if properly carried out.

7. "The Royal Government proceeded, on the very evening of the delivery of the note, to arrest Major Voislav Tankossitch. As regards Milan Ziganovitch, who is a subject of the Austro-Hungarian Monarchy, and who up to the 28th of June was employed (on probation) by the directorate of railways, it has not yet been possible to arrest him.

"The Austro-Hungarian Government is requested to be so good as to supply as soon as possible, in the customary form, the presumptive evidence of guilt, as well as the eventual proofs of their guilt, which have been collected up to the present in the enquiry at Sarajevo for the purposes of the later enquiry."

Austria-Hungary represented that Ciganovic did not leave Belgrade until three days after the Sarajevo outrage, and then by order of the police prefect, who declared afterwards in an interview that a man by the name of Milan Ciganovic did not exist in Belgrade, and that therefore the Serbian reply was disingenuous.

But this scarcely seems to be the inevitable conclusion from the circumstances cited by Austria-Hungary.

8. "The Serbian Government will reinforce and extend the measures which have been taken for preventing the illicit traffic of arms and explosives across the frontier. It goes without saying that they will immediately order an enquiry and will severely punish the frontier officials on the Schabatz-

Loznitza line who have failed in their duty and allowed the authors of the crime of Sarajevo to pass."

9. "The Royal Government will gladly give explanations of the remarks made by their officials, whether in Serbia or abroad, in interviews after the crime which, according to the statement of the Imperial and Royal Government, were hostile towards the Monarchy, as soon as the Imperial and Royal Government has communicated to them the passages in question in these remarks, and as soon as they have shown that the remarks were actually made by the said officials, although the Royal Government will itself take steps to collect evidence and proofs."

Austria-Hungary maintained that the Serbian government must have been aware of the interviews in question. The Serbian request for details and the reservation of the right to make a formal investigation are proof of the lack of a genuine intention of fulfilling the demand.

10. "The Royal Government will inform the Imperial and Royal Government of the execution of the measures comprised under the above heads, in so far as this has not already been done by the present note, as soon as each measure has been ordered and carried out."

"If the Imperial and Royal Government is not satisfied with this reply, the Serbian Government, considering that it is not to the common interest to precipitate the solution of this question, is ready, as always, to accept a pacific understanding, either by referring this question to the decision of the International Tribunal of the Hague, or to the Great Powers which took part in the drawing up of the declaration made by the Serbian Government on the 31st of March, 1909."

Austria-Hungary took no notice of the suggestion for an arbitration of any remaining unsatisfactory points by the Hague Tribunal, but with her declaration that the Serbian reply was a mere playing for time, she refused to accept it. Baron Giesl, the Austrian Minister, formally notified the Serbian government, at 6.30, the same afternoon, that the response had been delivered, and that he was about to depart from Belgrade with the personnel of the legation, because he had not received a satisfactory reply within the specified time. The Serbian government and foreign diplomatic corps withdrew the same evening to Nish, which became the temporary capital of the country.

M. Sazonoff informed the Russian representatives in France, Great Britain, Germany, Austria-Hungary, and Italy, on the 27th, that in his opinion the Serbian reply exceeded all expectations in moderation and the proof which it afforded of a desire to give the most complete satisfaction to Austria-Hungary; and that he could not understand what further demands Austria-Hungary could make, unless she were actually searching a pretext for war with Serbia.

Sir Edward Grey had received information, July 25, "that the Austrian method of procedure on expiry of the time limit would be to break off diplomatic relations and commence military preparations, but not military operations." He remarked to the German ambassador that this would interpose a stage of mobilization before the frontier was actually crossed, and therefore a further opportunity for negotiations. He believed, therefore, that the one chance of peace would be for Italy, France, Russia, and Great Britain to keep together and to join in asking Austria and Russia not to commence hostile action until they had had time to arrange matters. This idea of common action by the four powers took the practical form of a proposal for a conference to be composed of Sir Edward Grey and the ambassadors, as embodied in the following message sent to the British diplomatic representatives in Paris, Berlin, and Rome, July 26:

"Would minister for foreign affairs be disposed to instruct ambassador here to join with representatives of France, Italy, and Germany, and myself to meet here in conference immediately for the purpose of discovering an issue which would prevent complications? You should ask minister for foreign affairs whether he would do this. If so, when bringing the above suggestion to the notice of the governments to which they are accredited, representatives at Belgrade, Vienna, and St. Petersburg should be authorized to request that all active military operations should be suspended pending results of conference."

The French and Italian governments immediately accepted the proposal, and on the 27th the Russian government signified its approval of the conference in case direct explanations with Austria-Hungary were impossible. It must be explained that on the day previous M. Sazonoff had entered into a conversation with the Austro-Hungarian ambassador, which had been carried on in a friendly tone, and that he had transmitted a request to Vienna that a continuation of this exchange of views should be permitted.

But Sir Edward Goschen, British Ambassador at Berlin, was obliged to report Germany's refusal to fall in with the idea of a conference of this kind:

"Secretary of State (for Foreign Affairs) says that conference you suggest would practically amount to a court of arbitration, and could not, in his opinion, be called together except at the request of Austria and Russia. He could not, therefore, fall in with your suggestion." The foreign secretary added that the news from St. Petersburg revealed an intention on the part of M. Sazonoff to exchange views with Count Berchtold, and it would be best to await the outcome.

The refusal to take part in this conference has exposed the foreign policy of Germany to very severe criticism. It can scarcely be supposed that this refusal was due solely to a chivalrous consideration for the feelings of her ally. It may be regarded, no doubt, as part of Germany's policy of scrupulously avoiding every appearance of interfering in the Austro-Serbian controversy, or recognizing by any act or statement that it was an affair which concerned any other states.

The Chancellor telegraphed to Prince Lichnowsky, the German ambassador in London, July 27, that it was impossible for Germany to place her ally Austria-Hungary in the latter's dispute with Serbia before a European tribunal, and that Germany's mediation must be limited to the danger of an Austro-Russian conflict.

Later, the same day, he sent another communication to Prince Lichnowsky stating that he had started in Vienna the proposal for mediation between Austria-Hungary and Russia in the sense desired by Sir Edward Grey; but on the following day he reported that Count Berchtold found the mediation proposal with Russia belated because hostilities with Serbia had already begun. Evidently Count Berchtold did not observe the rather fictitious distinction between an Austro-Serbian and an Austro-Russian controversy as defined by the Chancellor.

The Kaiser returned unexpectedly to Germany from his Norwegian cruise, July 26. It was generally recognized that he had exerted his personal influence, probably with decisive results, in favor of the maintenance of peace at the termination of the Balkan crisis the summer before. It is a problem of great interest and importance to endeavor to penetrate the nature of his influence upon the foreign relations of Germany during those days of extreme tension which directly preceded the war. On the same day the Russian Foreign Minister Sazonoff had had a long conversation in a friendly tone with the Austrian Ambassador Szapory regarding the contents of the Austro-Hungarian note. This led him to transmit through the Russian ambassador in Vienna the request that the Austro-Hungarian ambassador in St. Petersburg should be authorized to carry on an exchange of views with him relative to the revision of the note by finding a formula acceptable to Serbia and at the same time guaranteeing satisfaction to Austria-Hungary. But the Russian ambassador replied the next day that Count Berchtold was again absent from Vienna so that he had not been able to obtain a reply to this message.

A confidential communication sent by the Chancellor to the different state governments of Germany to explain the policy of the imperial government, July 28, at the same time that Austria-Hungary declared war against Serbia, gives no indication of change in Germany's attitude. After reciting the features of the situation as attested by the Austro-Hungarian note to Serbia, the connivance of Serbian officers and officials in the fatal agitation leading to the outrage of Sarajevo, the moderation of Austria-Hungary in the past in the face of Serbian chauvinism, the failure of Serbia to keep her promises, and the absolute necessity for Austria-Hungary to assume a vigorous attitude, the letter continued as follows:

"The reply of the Serbian government to the demands which the Austro-Hungarian government put on the 23rd inst. through its representative in Belgrade, shows that the dominant factors in Serbia are not inclined to cease their former policies and agitation. There will remain nothing else for the Austro-Hungarian government than to press its demands, if need be through military action, unless it renounces for good its position as a great power.

"Some Russian personalities deem it their right as a matter of course and a task of Russia's activity to become a party to Serbia in the conflict between Austria-Hungary and Serbia. For the European conflagration which would result from such a step by Russia, the "*Novoye Vremya*" believes itself justified in making Germany responsible in so far as she does not induce Austria-Hungary to yield.

"The Russian press thus turns conditions upside down. It is not Austria-Hungary which has called forth the conflict with Serbia, but it is Serbia which, through unscrupulous favor toward Pan-Serbian aspirations, even in parts of the Austro-Hungarian Monarchy, threatens the same in her existence and creates conditions, which eventually found expression in the wanton outrage at Sarajevo. If Russia believes that it must champion the cause of Serbia in this matter, it certainly has the right to do so. However, it must realize that it makes the Serbian activities its own, to undermine the conditions of existence of the Austro-Hungarian Monarchy, and that thus it bears the sole responsibility if out of the Austro-Serbian affair, which all the other great powers want to localize, there arises a European war. This responsibility of Russia's is evident and it weighs the more heavily as Count Berchtold has officially declared to Russia that Austria-Hungary has no intention to acquire Serbian territory or to touch the existence of the Serbian Kingdom, but only desires peace against the Serbian intrigues threatening its existence.

"The attitude of the Imperial Government in this question is clearly indicated. The agitation conducted by the Pan-Slavs in Austria-Hungary has for its goal, with the destruction of the Austro-Hungarian Monarchy, the scattering or weakening of the Triple Alliance, with a complete isolation of the German Empire in consequence. Our own interest therefore calls us to the side of Austria-Hungary. The duty, if at all possible, to guard Europe against a universal war, points to a support by ourselves of those endeavors which aim at the localization of the conflict, faithful to the course of those policies which we have carried out successfully for forty-four years in the interest of the preservation of the peace of Europe.

"Should, however, against our hope, through the interference of Russia the fire be spread, we should have to support, faithful to our duty as allies, the neighbor-monarchy with all the power at our command. We shall take the sword only if forced to it, but then in the clear consciousness that we are not guilty of the calamity which war will bring upon the peoples of Europe."

This is the German position stated in the clearest terms, and there is nothing to prove that the Teutonic powers altered their attitude at any time during the discussion. As the German government viewed the situation after the declaration of hostilities by Austria-Hungary against Serbia, the question remained exclusively one between these two powers, and localizing the struggle, and thus guarding Europe against a general war, was possible only by allowing the Dual Monarchy a free hand. To prevent the intervention of a third power, as Russia, Germany was prepared to support her ally with all her forces. The attitude of Germany and that of Russia on this point were absolutely contradictory, and whichever view seems the more reasonable or just, it is clear that the German position permitted very little room for discussion with Russia. The only practical field left open for negotiation in harmony with this view lay in convincing Russia of the sincerity of Austria-Hungary's formal repudiation of any intention of annexing Serbian territory or disturbing the balance of power in the Balkan peninsula.

The Kaiser took the initiative in a direct exchange of telegrams with the Tsar, reminding him on the 28th that they, together with all other sovereigns, had a common interest in securing the punishment of all those who were responsible for the horrible murder of Archduke Francis

Ferdinand, and promising to use his influence to induce Austria-Hungary to obtain a frank and satisfactory understanding with Russia.

The Tsar replied on the 29th:

"I am glad that you are back in Germany. In this serious moment I ask you earnestly to help me. An ignominious war has been declared against a weak country and in Russia the indignation which I fully share is tremendous. I fear that very soon I shall be unable to resist the pressure exercised upon me and that I shall be forced to take measures which will lead to war. To prevent such a calamity as a European war would be, I urge you in the name of our old friendship to do all in your power to restrain your ally from going too far."

The Kaiser in his turn telegraphed toward evening, the same day:

"I have received your telegram and I share your desire for the conservation of peace. However; I cannot—as I told you in my first telegram—consider the action of Austria-Hungary as an 'ignominious war.' Austria-Hungary knows from experience that the promises of Serbia as long as they are merely on paper are entirely unreliable. According to my opinion the action of Austria-Hungary is to be considered as an attempt to receive full guaranty that the promises of Serbia are effectively translated into deeds. In this opinion I am strengthened by the explanation of the Austrian cabinet that Austria-Hungary intended no territorial gain at the expense of Serbia. I am therefore of opinion that it is perfectly possible for Russia to remain a spectator in the Austro-Serbian war without drawing Europe into the most terrible war it has ever seen. I believe that a direct understanding is possible and desirable between your government and Vienna, an understanding which—as I have already telegraphed you—my government endeavors to aid with all possible effort. Naturally military measures by Russia, which might be construed as a menace by Austria-Hungary, would accelerate a calamity which both of us desire to avoid and would undermine my position as mediator which—upon your appeal to my friendship and aid—I willingly accepted."

Although this exchange of telegrams continued until the 31st, the Kaiser complained that his action as mediator was rendered increasingly difficult by the Russian mobilization.

In a council of the Russian ministers held in the presence of the Tsar, July 25, a measure was adopted to provide for an eventual mobilization of thirteen army corps destined to operate against Austria-Hungary, which was to be carried out only in case Austria-Hungary should attack Serbia, and at the discretion of the Russian minister of foreign affairs. On the 27th the secretary of war gave the German military *attaché* his word of honor that no order to mobilize had as yet been issued, and that, although general preparations were being made, no reserves had been called and no horses mustered. He added that if Austria crossed the Serbian frontier such military districts as are directed towards Austria-Hungary; Kieff, Odessa, Moscow, and Kazan, would be mobilized, but not those, Warsaw, Wilna, and St. Petersburg, that faced towards the German frontier. This statement is in accordance with the resolutions of the ministers on July 25th.

Austria-Hungary had partially mobilized on the 26th, and declared war against Serbia on the 28th. This action was reflected in the increased state of international tension the next day. All the nations seem to have advanced in the direction of war on the 29th. The immediate effect of the declaration of hostilities was Russia's partial mobilization in accordance with the cabinet decision of the 25th, in view of the evident absence of a desire on Austria's part "to accept any kind of peaceful solution of her conflict with Serbia." The declaration of war by Austria-Hungary had been assigned by Count Berchtold as obstacle to any consideration of mediation proposals. Upon the return of President Poincaré and Prime Minister Viviani to Paris on July 29, the latter confirmed the assurance of French solidarity and complete accord with Russia.

A telegram of Sir M. de Bunsen, British ambassador at Vienna, to Sir Edward Grey on the 30th contains an interesting statement regarding the purpose of the Russian mobilization and the attitude of Russia at this critical time:

"Russian ambassador hopes that Russian mobilization will be regarded by Austria as what it is, *viz.;* a clear intimation that Russia must be consulted regarding the fate of Serbia, but he does not know how the Austrian government are taking it. He says that Russia must have an assurance that Serbia will not be crushed, but she would understand that Austria-Hungary is compelled to exact from Serbia measures which will secure her Slav provinces from the continuance of hostile propaganda from Serbian territory. The French ambassador hears from Berlin that the German ambassador at Vienna (Herr von Tschirscky) is instructed to speak seriously to the Austro-Hungarian government against acting in a manner calculated to provoke a European war. Unfortunately the German ambassador is himself so identified with extreme anti-Russian and anti-Serbian feeling prevalent in Vienna that he is unlikely to plead the cause of peace with entire sincerity. Although I am not able to verify it, I have private information that the German ambassador knew the text of the Austrian ultimatum to Serbia before it was despatched and telegraphed it to the German Emperor. I know from the German ambassador himself that he endorses every line of it."

On the 29th Count Pourtalès, German Ambassador at St. Petersburg, told Foreign Minster Sazonoff that the German government was willing to guarantee that Serbian integrity would be respected by Austria. But M. Sazonoff objected that even under those circumstances Serbia might become an Austrian vassal, and that there would be a revolution in Russia, if the Russian government should tolerate such a state of affairs.

After the partial mobilization in Russia had been reported in Germany, the Kaiser telegraphed to the Tsar at one o'clock in the morning of the 30th to emphasize the dangerous consequences of Russian mobilization, and to warn him that it would threaten, if not render impossible, his rôle of mediator.

At two o'clock the same morning, the German ambassador called upon M. Sazonoff again, in a despairing effort to preserve peace, and begged him to make some suggestion which he could telegraph to the German government as a last hope. Accordingly M. Sazonoff drew up the following formula:

"If Austria, recognizing that her conflict with Serbia has assumed the character of a question of European interest, declares herself ready to eliminate from her ultimatum the points which violate the principle of the sovereignty of Serbia, Russia engages to stop all military preparations."

M. Sazonoff already claimed to have absolute proof that Germany was making active military and naval preparations against Russia. It was now recognized that if Austria-Hungary did not accept the formula proposed by M. Sazonoff, Russia would proceed to a general mobilization, and a European war would be the inevitable result. On the same day the German Secretary for Foreign Affairs, Herr von Jagow, informed the Russian ambassador that he considered the Russian proposal inacceptable.

On the 31st, M. Sazonoff modified his formula, at the suggestion of Great Britain, to the following form:

Pierre Paul Cambon, French Ambassador to Great Britain.

Gottlieb von Jagow, Secretary of State for Foreign Affairs of the German Empire.

"If Austria will agree to check the advance of her troops on Serbian territory; if, recognizing that the dispute between Austria and Serbia has assumed a character of European interest, she will allow the Great Powers to look into the matter and determine whether Serbia could satisfy the Austro-Hungarian government without impairing her own rights as a sovereign state or her independence, Russia will undertake to maintain her waiting attitude."

This final concession was too late to produce any results, and it is by no means certain that Austria-Hungary would have been willing at any time to admit that her quarrel with Serbia was a question of European interest. Events were being impelled irresistibly towards a catastrophe. On the 31st the Russian government ordered a general mobilization, in consequence, probably, of the reported unyielding attitude of Austria-Hungary, the announcement of a general mobilization in that country, and supposed measures of military preparation taken by Germany. It had been generally recognized from the first that general mobilization by Russia would lead almost immediately to the same measure by Germany.

Accordingly, the Chancellor sent the following telegram to the German ambassador in St. Petersburg, instructing him to deliver an ultimatum to M. Sazonoff:

"In spite of negotiations still pending, and although we have up to this hour made no preparations for mobilization, Russia has mobilized her entire army and navy, hence also against us. On account of these Russian measures we have been forced, for the safety of the country, to proclaim the threatening state of war, which does not yet imply mobilization. Mobilization, however, is bound to follow if Russia does not stop every measure of war against us and against Austria-Hungary within twelve hours and notifies us definitely to this effect. Please to communicate this at once to M. Sazonoff and wire hour of communication."

As was to be foreseen, Russia did not comply with this demand, and on August 1, at 12.52 p. m., the Chancellor telegraphed to the German ambassador in St. Petersburg, instructing him to deliver the following statement, if by five o'clock the Russian government had not given a satisfactory reply to the ultimatum demanding demobilization:

"The Imperial Government has employed every effort from the beginning of the crisis to conduct matters to a peaceful solution. In compliance with the desire expressed to him by His Majesty the Emperor of Russia, His Majesty the Emperor of Germany, in accord with England, had undertaken the part of mediator between the cabinets of Vienna and St. Petersburg, when Russia without awaiting the result proceeded to mobilize her entire forces on land and sea. In consequence of this threatening measure, for which there had been no military preparation by Germany as justification, the German Empire found itself exposed to very serious, imminent danger. If the Imperial Government had failed to safeguard its interests against this peril, it would have compromised the safety, and even the existence, of Germany. Consequently, the German Government found itself under the necessity of addressing the government of His Majesty, the Emperor of all the Russias, to insist that the said military measures should cease. Since Russia has refused to comply with this demand, and has shown by this refusal that her action was directed against Germany, I have the honor, by order of my government, to announce the following to your Excellency:

'His Majesty, the Emperor, my august Sovereign, in the name of the empire, accepts the challenge and considers himself in a state of war with Russia'."

This declaration was actually delivered to M. Sazonoff at 7.10 p. m. At the same time that he transmitted the

German ultimatum to Russia on the 31st, the Chancellor despatched the following message to the German ambassador in Paris:

"Russia has ordered mobilization of her entire army and fleet, therefore also against us in spite of our still pending mediation. We have therefore declared the threatening state of war which is bound to be followed by mobilization unless Russia stops within twelve hours all measures of war against us and Austria. Mobilization inevitably implies war. Please ask French government whether it intends to remain neutral in a Russo-German war. Reply must be made in eighteen hours. Wire at once hour of inquiry. Utmost speed necessary."

On the following day, August 1, at 1.05 p.m., Herr von Schön, the German Ambassador in Paris, replied:

"Upon my repeated definite inquiry whether France would remain neutral in the event of a Russo-German war, the prime minister declared that France would do that which her interests dictated."

On the 3rd at 6.45 p. m. the German ambassador presented the following note, embodying Germany's declaration of war to M. René Viviani, Prime Minister and Minister of Foreign Affairs of France:

"The German civil and military authorities have observed a number of unmistakably hostile acts committed by French military aviators on German territory. Many of the latter have manifestly violated the neutrality of Belgium by flying over that country; one has attempted to destroy some buildings near Wesel, others have been sighted over the Eiffel region, another has dropped bombs on the railway near Karlsruhe and Nuremberg.

"I am directed, and have the honor, to inform Your Excellency that in presence of these aggressive acts the German Empire considers itself in a state of war with France, by the fault of this latter power.

"I have likewise the honor to inform Your Excellency that the German authorities will retain all French merchant ships in the German ports; but will release them, if a complete reciprocity of treatment is assured within forty-eight hours..........."

During the discussions preceding the war the Kaiser made repeated reference to his mediating influence at Vienna; but as the correspondence which passed between the German and Austro-Hungarian governments has not been published, there is no positive evidence for testing the true nature of the Kaiser's activity in this direction. He appears to have been sincerely desirous of preserving peace. It is commonly believed that the Vienna government adopted a conciliatory policy at the last moment, when it was certain that otherwise a general European war could not be avoided, and that they expressed a willingness to treat with Russia on the basis of the contents of their note to Serbia. Such a sudden revolution in their attitude is ascribed by one group of apologists to the Kaiser's beneficent influence, but by the other group to Austria-Hungary's timidity or apprehensions and conversely, the responsibility for the sudden interruption of this auspicious tendency is assigned by one party to Russia's mobilization, by the other, to the unwarranted German ultimatum and subsequent declaration of war. In view of the contents of the circular letter sent by the German government to the federated states, July 28, and the fact that the Kaiser makes no corresponding allusion in his telegrams, it seems very doubtful whether his conciliatory advice to Austria-Hungary went so far as to advocate a discussion with Russia, or the powers, on the basis of the demands in the note to Serbia, the only way in which his influence might have been productive of positive results.

The opinion that Austria-Hungary practically yielded at

Count Leopold von Berchtold, Minister of Foreign Affairs, Austria-Hungary.

Serge Sazonoff, Minister of Foreign Affairs, Russia.

the last is supported by the following evidence. Conversations between Vienna and St. Petersburg were resumed on July 30, and this may very likely have been partly due to the influence of the German Kaiser. The Russian ambassador in Paris informed the Russian foreign office on August 1 that the Austro-Hungarian ambassador in the French capital had declared that his government was willing to discuss the substance of its note to Serbia with the other powers. Furthermore, Sir M. de Bunsen, British Ambassador in Vienna, reported to his government that he was informed on August 1 by M. Schébéko that the Austrian ambassador in conversation with the Russian foreign minister in St. Petersburg had at last conceded the main point at issue by announcing that Austria-Hungary would submit to mediation the points in her note to Serbia which were incompatible with Serbian independence. But on the other hand, Count Berchtold informed M. Schébéko that Austria-Hungary could no longer recede or enter into any discussion of the terms of the note, July 30, and the next day M. Schébéko telegraphed the Russian foreign office that Count Berchtold insisted that Austria-Hungary must carry through her policy in regard to Serbia so as to secure an enduring guarantee of peace.

There is no official document to prove that Austria-Hungary yielded, and the Vienna "*Fremdenblatt*," an inspired sheet stated positively that the Vienna government did not change the direction of its policy to this extent.

CHAPTER VII

Great Britain and the War

Sir E. Grey's final efforts for peace. Germany's proposal for British neutrality, July 29. Belgian neutrality; guaranteed by the powers in 1839. Belgian neutrality in 1870; the temporary treaties; Mr. Gladstone's speech. Treaty of 1839 still binding. Belgium and the Congo; both were neutralized territory. Great Britain protests against Germany's demand for free passage across Belgium. Ultimatum and declaration of war, August 4. German views of causes and motives for Great Britain's intervention. Chancellor von Bethmann-Hollweg's speech, December 2, 1914, on Great Britain's position with respect to the war. Partial obligation for Great Britain to help France. Other possible causes of Great Britain's declaration of hostilities; commercial rivalry and alarm at development of German naval power. Commercial rivalry not a direct cause.

After the declaration of hostilities by Austria-Hungary against Serbia, July 28, M. Sazonoff informed London that this step clearly put an end to the idea of direct communications between Austria and Russia, and that action by the British cabinet became most urgent in order to obtain a suspension of military operations; for, otherwise, mediation would only allow matters to drag on and give Austria time to crush Serbia.

On the 29th, Sir Edward Grey told the German ambassador in London that, while he believed that a direct agreement between Russia and Austria-Hungary would be the best possible solution, he had just received information that the Austrian government had declined the suggestion of the Russian government that the Austrian ambassador at St. Petersburg should be authorized to discuss with M. Sazonoff the means of settling the Austro-Serbian conflict. He understood that the German government had said that they were favorable in principle to mediation between Austria and Russia, but objected to the form in which it had been proposed. He urged, therefore, "that the Cerman government should suggest any method by which the influence of the four powers could be used together to prevent war between Austria and Russia. France agreed, Italy agreed. The whole idea of mediation or mediating influence was ready to be put into operation by any method that Germany could suggest." Germany had only to "press the button" in the interests of peace. On the 30th, Herr von Jagow reported that he had asked the Austro-Hungarian government what would satisfy them, but had received no answer. The suggestion was soon lost in the turmoil of events.

About midnight on the 29th, Sir Edward Grey received an ominous communication from Sir E. Goschen, British Ambassador in Berlin, throwing a startling light on the extent to which the menace of war had already penetrated. Sir E. Goschen had been summoned for an interview with the Chancellor the same evening, after the latter had returned from Potsdam. The Chancellor pointed out that in spite of his continued efforts to maintain peace, a general war would become inevitable if Russia attacked Austria. He proceeded to bid for British neutrality by declaring that Germany was prepared to offer every assurance not to make any territorial acquisitions at the expense of France in case of victory, provided the neutrality of Great Britain were certain. Sir E. Goschen inquired about the French colonies, but the Chancellor was unable to make the same promise with regard to them. Germany would respect the neutrality of Holland, and "it depended upon the action of France what operations Germany might be forced to enter upon in Belgium, but when the war was over, Belgian integrity would be respected, if she had not sided against Germany." The Chancellor continued to say that in harmony with his policy of a better understanding with Great Britain he had in mind a general neutrality agreement between the two countries, and British neutrality in the impending conflict "would enable him to look forward to the realization of his desire." Sir E. Goschen replied that he thought Sir Edward Grey would desire to retain full liberty of action.

In reference to the Chancellor's proposals, Sir Edward Grey replied to Sir E. Goschen's communication the next day, July 30:

"His Majesty's government cannot for a moment entertain the Chancellor's proposal that they should bind themselves to neutrality on such terms.

"What he asks us in effect is to engage to stand by while French colonies are taken and France is beaten, so long as Germany does not take French territory as distinct from the colonies.

"From the material point of view such a proposal is unacceptable, for France, without further territory in Europe being taken from her, could be so crushed as to lose her position as a great power, and become subordinate to German policy.

"Altogether apart from that, it would be a disgrace for us to make this bargain with Germany at the expense of France, a disgrace from which the good name of this country would never recover.

"The Chancellor also in effect asks us to bargain away whatever obligation or interest we have as regards the neutrality of Belgium. We could not entertain that bargain either."

Sir Edward Grey went on to urge that the one way for improving the relations between Great Britain and Germany was for the two countries to work together to preserve the peace of Europe in the existing crisis.

Sir Edward Grey continued to strive manfully for peace up to the very last, refusing to give up hope even when the *ultimata* were being presented. Thus, on the 31st he told the German ambassador that if "Germany could get any reasonable proposal put forward which made it clear that Germany and Austria were striving to preserve European peace, and that Russia and France would be unreasonable if they rejected it, he would support it at St. Petersburg and Paris, and go the length of saying that if Russia and France would not accept it His Majesty's government would have nothing more to do with the consequences." He told him that otherwise, if France became involved, they would be drawn in, too. Herr von Jagow expressed sympathy for Sir Edward Grey's continued efforts to maintain peace, but said that until they received an answer from their demand that Russia should demobilize, it was impossible to consider any proposal.

On the same day Sir Edward Grey addressed identical inquiries to the French and German governments through the following message sent to the British ambassadors at Paris and Berlin respectively:

"I still trust that situation is not irretrievable, but in view of prospect of mobilization in Germany it becomes essential to His Majesty's government, in view of existing treaties, to ask whether French (German) government are prepared to engage to respect neutrality of Belgium so long as no other power violates it.

"A similar request is being addressed to German (French) government. It is important to have an early answer."

The French government replied that they were resolved to respect Belgian neutrality unless it were violated by some other power. But Herr von Jagow told the British ambassador at Berlin that he must first consult with the Kaiser and Chancellor, and he was very doubtful whether they would return any answer, since any reply might disclose a part of their plan of campaign.

On the next day, August 1, Sir Edward Grey expressed to the German ambassador his regret at the nature of the reply of the German government, pointing out how keenly the neutrality of Belgium affected feeling in Great Britain. Prince Lichnowsky inquired whether Great Britain would engage to remain neutral if Germany promised not to violate Belgian neutrality. Sir Edward Grey could only say that the neutrality of Belgium would appeal very strongly to public opinion, which would largely determine the attitude of the government. When the German ambassador asked further whether he could not formulate conditions under which they would remain neutral, he said that he "felt obliged to refuse definitely any promise to remain neutral on similar terms." He could only say that they must keep their hands free.

The neutrality of Belgium as guaranteed by the Great Powers is a factor of such fundamental importance among the causes of the war that a reference at some length to its origin, and the previous attitude of the powers with respect to it, is appropriate.

The Congress of Vienna in 1815 united what is now Belgium with Holland to constitute the Kingdom of the Netherlands under the rule of the House of Orange. The congress intended to create in this way a stronger buffer state, so as to restrict the ambition for aggrandizement of France, or any other power. But the populations of the two parts of the new kingdom were too dissimilar to combine harmoniously into a united national state. The inflexible attitude of the government, which preserved its essentially Dutch character, aggravated the inevitable discord, and resulted in the revolution in Belgium in 1830, which broke up the union. The independence of Belgium was finally recognized by Holland in 1839.

Two treaties signed on the same day, April 19, 1839, in

London, were the basis of Belgian neutralization. Belgium and Holland were the parties to one of these treaties. It sanctioned their separation and individual independence, and contained the following provision:

"Belgium in the limits above described shall form a state independent and perpetually neutral. She shall be bound to observe this same neutrality toward all other states."

Five powers were parties to the second treaty, Great Britain, France, Prussia, Austria, and Russia, and for this reason it is called the Quintuple Treaty. By it the powers mentioned approved and adopted for themselves the provisions of the first treaty. A little while later the German Confederation also became a party to this treaty.

The outbreak of hostilities between France and the North German Federation in July, 1870, emphasized the importance of Belgian neutrality. Early in August, the British Parliament was very much aroused by Bismarck's disclosure of the draft of a proposed agreement between France and Prussia drawn up three years earlier to sanction the eventual annexation of Belgium by France under certain circumstances. Prussia had finally refused to ratify this agreement. With the startling impression of this revelation still fresh, Gladstone ascribed such great importance to the preservation of Belgian neutrality that he secured separate but identical treaties from the North German Federation and France, on the 9th and 26th of August respectively, guaranteeing the neutrality of Belgium during the period of the war and for a year after its termination. The third article of these treaties is of fundamental importance in its bearing upon the question of the continuous validity of the original treaty of 1839. It reads as follows:

"This treaty shall be binding on the high contracting parties during the continuance of the present war between France and the North German Federation, and for twelve months after the ratification of any treaty of peace concluded between these parties; and on the expiration of that time the independence and neutrality of Belgium will, so far as the high contracting parties respectively are concerned, continue to rest as heretofore on the first article of the Quintuple Treaty of the 19th of April, 1839."

When the signing of the first of these treaties had been announced to parliament, it was criticised by some who expressed the opinion that it was an unnecessary addition to, or substitution for, the earlier treaty, and by others who were afraid that it might involve Great Britain needlessly in the war. Accordingly, on August 10, Mr. Gladstone made a statement in the House of Commons regarding the purpose of these treaties, and a portion of his remarks will show us the attitude of the British government at that time with regard to Belgium, and reveal in particular their interpretation of the situation with respect to the guaranteed neutrality of that country:

"There is, I admit, the obligation of the treaty (of 1839). It is not necessary, nor would time permit me, to enter into the complicated question of the nature of the obligations of that treaty; but I am not able to subscribe to the doctrine of those who have held in this house what plainly amounts to an assertion that the simple fact of the existence of a guarantee is binding on every party to it irrespective altogether of the particular position in which it may find itself at the time when the occasion for acting on the guarantee arises. The great authorities on foreign policy to whom I have been accustomed to listen—such as Lord Aberdeen and Lord Palmerston—never, to my knowledge, took that rigid and, if I may venture to say so, that impracticable view of a guarantee. The circumstance that there is already an existing guarantee in force is of necessity an important fact, and a weighty element in the case, to which we are bound to give

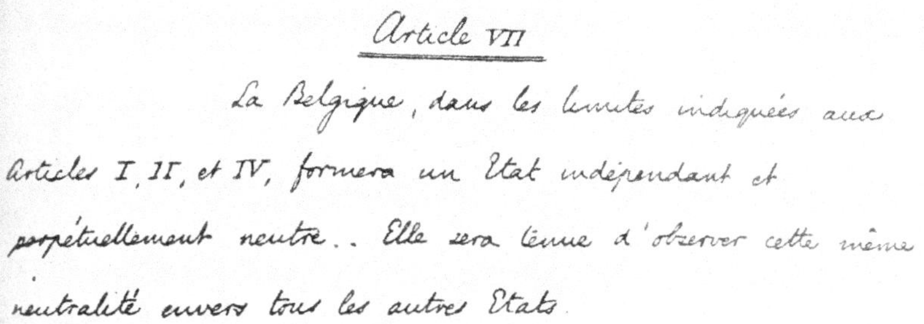

Article VII

La Belgique, dans les limites indiquées aux Articles I, II, et IV, formera un Etat indépendant et perpétuellement neutre. Elle sera tenue d'observer cette même neutralité envers tous les autres Etats.

Article VII of the treaty of 1839 which guaranteed the independence and perpetual neutrality of Belgium.

Last page of the treaty of 1839 which guaranteed the neutrality of Belgium, showing the als and signatures of the plenipotentiaries of Austria, Great Britain, Belgium, France, Russia, d Prussia.

full and ample consideration. There is also this further consideration, the force of which we must all feel most deeply and that is common interest against the unmeasured aggrandizement of any power whatsoever.

"There is one other motive, which I shall place at the head of all, that attaches peculiarly to the preservation of the independence of Belgium. What is that country? It is a country containing 4,000,000 to 5,000,000 of people, with much of an historic past, and imbued with a sentiment of nationality and a spirit of independence as warm and as genuine as that which beats in the hearts of the proudest and most powerful nations. By the regulation of its internal concerns, amid the shocks of revolution, Belgium, through all the crises of the age, has set to Europe an example of a good and stable government gracefully associated with the widest possible extension of the liberty of the people. Looking at a country such as this, is there any man who hears me who does not feel that if, in order to satisfy a greedy appetite for aggrandizement, coming whence it may, Belgium were absorbed, the day that witnessed that absorption would hear the knell of public right and public law in Europe?

"But we have an interest in the independence of Belgium which is wider than that—which is wider than that which we may have in the literal operation of the guarantee. It is found in the answer to the question whether, under the circumstances of the case, this country, endowed as it is with influence and power, would quietly stand by and witness the perpetration of the direst crime that ever stained the pages of history, and thus become participators in the sin?

"But in what, then, lies the difference between the two treaties? It is in this—that, in accordance with our obligations, we should have had to act under the treaty of 1839, without stipulated assurance of being supported from any quarter whatever, against any combination however formidable; whereas by the treaty now formally before parliament, under the conditions laid down in it, we secure powerful support in the event of our having to act..........

"It is perfectly true that this is a cumulative treaty, added to the treaty of 1839, as the right honorable gentleman opposite, with perfect precision, described it........

"The treaty of 1839 loses nothing of its force even during the existence of this present treaty. There is no derogation from it whatever. The treaty of 1839 includes terms which are expressly included in the present instrument, lest by any chance it should be said that in consequence of the existence of this instrument, the treaty of 1839 had been injured or impaired."

The North German Federation, as we have seen, formally accepted, or confirmed, the obligation in regard to Belgian neutrality embodied in the treaty of 1839 by signing the compact with Great Britain, Aguust 9, 1870, in which it was distinctly stipulated that the provisions of the earlier treaty should remain in force. It would be mere casuistry, deceptive quibbling, to maintain that the treaty obligations of the North German Federation were annulled by its enlargement into the present German Empire. The renunciation of solemn engagements on this account would suggest the cheating of creditors by deceptive business reorganizations. Successors in business firms assume the liabilities as well as the assets.

When Germany adopted a threatening attitude towards France, in March, 1875, on account of the latter's rapid recovery of military strength, Bismarck inquired of the Belgian government what measures they proposed to take for preserving their neutrality. He may have had in mind only the general kind of neutrality of any non-combatant. However, the German Empire has never indicated by any act or declaration before the outbreak of the present war that it did not regard the Quintuple Treaty of 1839 as binding upon itself and the other signatory powers.

The political activity of Belgium has displayed a development during the last generation in a direction entirely unforeseen when Belgian territory was declared perpetually inviolable in 1839, a development which might seem to render the stipulated protection of the Great Powers less appropriate.

The Society for the Investigation of the Upper Congo was founded in Belgium under the protection of King Leopold II in 1878, and an exploration party was sent out to Central Africa with Mr. Henry M. Stanley at its head. By this means, the society, which became later the International Association of the Congo, established a claim by occupation to nearly all the vast area drained by the Congo River. Thus, on February 23, 1885, the association was able to announce at an international congress held in Berlin for the purpose of formulating the public law of Africa that it had obtained the recognition of its sovereignty in this region by all the powers. Accordingly, the Congo Free State was solemnly inaugurated, August 1, 1885, with Leopold II of Belgium as its king. This monarch had always looked forward to the annexation of the new state to Belgium, and prescribed this devolution in the terms of his will. Later, it was enacted in 1890 that the annexation should take place at the termination of ten years; but it was not finally consummated until 1909, when Belgium came into possession of a colonial area of 909,654 square miles with valuable natural resources.

The argument has been advanced that this extension of Belgium's sovereignty over a large colonial area is incompatible with her permanent neutrality; that a state which enters into political competition with other states has no right to be shielded from the results of full responsibility for its actions. But, on the other hand, the Congo Free State was declared neutral by act of the Berlin Congress, February 26, 1885, at the same time that it was opened freely to the trade of all nations, and there seems to be no logical difficulty in assuming that the acquisition of a neutral dependency by a neutral state has not diminished the neutrality of either. Besides, the motive for the neutralization of the Congo Free State was not unlike that which suggested the neutralization of Belgium. For the Congo territory occupies a central position in the heart of the continent between the possessions of the other powers; and at a time when the nations were eagerly extending their pretensions towards the complete partition of Africa, it was certainly a very important precautionary measure against possible conflicts to remove this large intervening space from the field of international rivalry. Therefore, since the neutrality of Belgium and the Congo are international interests in the respective continents in which they are situated, it is reasonable that they should be politically united, so that the first may be the advocate and support of the neutrality of the second.

On August 4, the British government received the following communication, addressed by the King of the Belgians to the British King:

"Remembering the numerous proofs of Your Majesty's friendship and that of your predecessor, and the friendly attitude of England in 1870, and the proof of friendship you have just given us again, I make a supreme appeal to the diplomatic intervention of Your Majesty's government to safeguard the integrity of Belgium."

In the morning of the same day, since news had been received that Germany had demanded a free passage across Belgian territory, threatening in the case of refusal to treat that country as an enemy, Sir Edward Grey instructed Sir

Sir Edward Grey, Secretary of State for Foreign Affairs, Great Britain.

E. Goschen, British Ambassador in Berlin to request an assurance that the demand would not be proceeded with and that Belgian neutrality should be respected. During the course of the day, information was received that the German minister in Brussels had notified the Belgian foreign office that the German government would be compelled to carry out their proposed measures by force; and then, that German troops had entered Belgian territory.

Accordingly Sir Edward Grey instructed the British minister in Brussels to inform the Belgian government that the British government expected them to resist by any means in their power German pressure to depart from neutrality and that it would support them in offering such resistance. Later in the same day, he sent the following memorable telegram to Sir E. Goschen:

"We hear that Germany has addressed note to Belgian minister for foreign affairs stating that German government will be compelled to carry out, if necessary, by force of arms, the measures considered indispensable. We are also informed that Belgian territory has been violated at Gemmenich.

"In these circumstances, and in view of the fact that Germany declined to give the same assurance respecting Belgium as France gave last week in reply to our request made simultaneously at Berlin and Paris, we must repeat that request, and ask that a satisfactory reply to it and to my telegram of this morning be received here by 12 o'clock tonight. If not, you are instructed to ask your passports, and to say that His Majesty's government feel bound to take all steps in their power to uphold the neutrality of Belgium and the observance of a treaty to which Germany is as much a party as ourselves."

To the representations of Sir E. Goschen, Herr von Jagow could only reply in the sense that the German forces had already entered Belgium and they could not withdraw. Accordingly, Great Britain's declaration of war went into effect.

The violation of the neutrality of Belgium by Germany was the professed cause for Great Britain's declaration of war against the latter power. But the Germans deny that this is a valid cause or represents the true motive for Great Britain's hostility. For, in the first place, they declare that they themselves were not under obligation to respect Belgium's neutrality, and in the second place, before the crisis occurred, Great Britain had practically pledged herself to support France against Germany, and she had even ceased to maintain an attitude of neutrality before Germany had trespassed on Belgium. The German view was formulated in a speech delivered by Chancellor von Bethmann-Hollweg before the Reichstag, December 2, 1914, on the first occasion for addressing them regarding the government's foreign policy since the commencement of hostilities with Great Britain, as follows:

"The apparent responsibility for the war rests on those individuals in Russia who demanded, and put into execution, the general mobilization of the Russian army. But the deeper responsibility must be attributed to the government of Great Britain. The London cabinet could have made the war impossible by declaring in unmistakable language in St. Petersburg that England would not permit the Austro-Serbian conflict to develop into a continental war of the Great Powers. Such an attitude would have compelled France to exert herself to restrain Russia from undertaking warlike measures. That would have insured the success of our efforts to mediate between Vienna and St. Petersburg, and there would have been no war. But England did not act in this way. England was aware of the bellicose intrigues of the partly irresponsible, but powerful, group among the Tsar's immediate courtiers. England perceived very clearly how the ball was rolling but put no obstacle in its path. In spite of all its peaceful protestations, the London government allowed it to be understood that England stood by France, and, therefore, by Russia. The Belgian *chargé d' affaires* in St. Petersburg sent the following report to his government, July 30:

" 'They are convinced in St. Petersburg, and even have definite assurance, that England will support France. The assurance of this support carries great weight and has contributed considerably to give the military party the upper hand'."

The Chancellor went on to explain that as late as the summer of 1914, the English ministers had affirmed that England was not bound by any treaty to participate in a European war, and derived the conclusion that "it follows necessarily that the London cabinet suffered this terrible world-war to develop because it seemed to offer them an opportunity, with the aid of their *entente* friends, of dealing a death-blow to their greatest competitor in the world-market.

"The Belgian neutrality which the English pretended to defend is a mask. On August 2, at 7 p. m., we announced in Brussels that the French plan of campaign, which was no secret to us, left us no alternative but to march across Belgium, as an act of self-preservation. But as early as the afternoon of the same day, the 2nd, that is, before London could have had any news of our action in Brussels, England had pledged her help to France—pledged unconditional assistance, in case a German fleet made an attack on the French coast. No mention was made of Belgian neutrality at that time. How can England maintain that she drew her sword because we violated Belgian neutrality?

"There were already many indications of the guilt of the

Belgian government. I had at that time no documentary evidence; but the English statesmen must have been very familiar with the proofs of Belgian complicity. Now that my publication of the documentary proof found in Brussels has revealed how, and to what extent, Belgium had already sacrificed her neutrality to England, the whole world is convinced of two facts.

"First: When our troops crossed the frontier of Belgium during the night of August 3-4, they entered the territory of a state which had given up its own neutrality long before.

"Second: The profession (by British government) of the violation of Belgian neutrality as cause for the declaration of war, August 4, was only a subterfuge to conceal the real motive from their own people, and neutral nations."

Some further remarks of a more general nature in regard to the Chancellor's policy during the years just passed may be added:

"We still encountered the spirit of revenge in France. Fostered by ambitious politicians, it proved to be more influential than the desire, which was undoubtedly cherished by a part of the French people, of living on good neighborly terms with us. We succeeded in coming to an understanding with Russia on some specific points; but Russia's unshakable alliance with France, her antagonism to our ally Austria-Hungary, and an anti-German prejudice nourished by Pan-Slavic greed for power, rendered impossible the establishment of an agreement firm enough to exclude the danger of war in periods of political crisis.

"England was relatively less hampered. There the attempt to effect an agreement capable of guaranteeing the peace of the world seemed hopeful."

"Our intercourse (with England) suffered another shock in the crisis of 1911. The English people realized over night that it had been brought to the verge of a European war. Popular feeling forced the government to adopt a more conciliatory attitude toward Germany. By wearisome efforts it was possible to arrive at an understanding regarding some economic points of controversy touching Asia Minor and Africa. In this way the potential sources of conflict were reduced. The world is large, and offers, unless there is a deliberate policy of hindering the development of our resources, room enough for both peoples to measure their strength in friendly competition."

These statements of the Chancellor embody the prevailing opinion in Germany regarding Great Britain's position with respect to the war, and we may conveniently employ them as a guide to direct our analysis to the indications of causes which are significant.

Our consideration of the history of Belgian neutrality has shown, not only that Great Britain had continuously regarded the guaranteed neutrality as valid, but that the German Empire was bound by the agreement, unless it presumed to repudiate the obligations sanctioned by the North German Federation, of whose personality it is the continuation.

But, as we have already seen, Chancellor von Bethmann-Hollweg declared in the course of his speech in the Reichstag, December 2, 1914, that Belgium had really ceased to be neutral before the German army entered her territory in the night of August 3-4, and this is the almost universal belief of the German people. Some documents, which were discovered in the war department in Brussels and first published in the *North German Gazette*, are the principal evidence for the German conviction that a secret agreement existed between Great Britain and Belgium directed against Germany, which invalidated Belgium's neutrality. The complicity of Belgium in a hostile plot against her powerful eastern neighbor has been proclaimed as justification of the forcible occupation of the country by the Germans. Germany was represented as the victim of a conspiracy, not an aggressor, and Belgium as a culprit which merited the misfortunes which she has suffered. According to this view, Great Britain, which posed as the disinterested protector of small nations, had really been the first to violate Belgium's neutrality by enticing her into a wicked plan for attacking Germany.

The first of these documents is a letter addressed by General Ducarme, the Belgian Chief of Staff to the Belgian Minister of War, dated April 10, 1906. The general reported in this communication a summary of a series of confidential conversations with Lieutenant Colonel Barnardiston, the British military *attaché*, which had been in progress since the middle of January. Ducarme and Barnardiston, at the latter's suggestion, had studied certain technical problems which would arise if Great Britain should send an army to aid in the defense of Belgium against Germany, such as railway transportation, the requisitioning of supplies, intermediary officers, and interpreters. The British *attaché* stated that Great Britain was prepared to despatch an expeditionary force of 100,000 men, and they would land at Dunkirk and Calais. General Ducarme declared that the Belgian field army of 100,000 men would be ready for action in four days. The document contained the specific statement of the British representative that the conversations had no binding character for the two countries. In a marginal addition the statement was made that the British intervention would not take place before Germany had violated Belgian neutrality.

The second document in the chronological order is doubtless a communication of the Belgian Ambassador in Berlin, Baron Greindl, of December 23, 1911, in which he warned his government that Great Britain and France proposed to make

Belgium a base for an aggressive action against Germany, and protested that a plan ought to be devised against a possible violation of their neutrality by the western powers.

The third and final document shows, that while the Belgian authorities may not have shared Baron Greindl's preoccupations, they were disposed to insist upon the correct observance of their neutrality by Great Britain. It has no formal date, but the text mentions April 23, and the year was probably 1912. The contents are similar to those of the first note. It has no signature. General Jungbluth had succeeded General Ducarme as Belgian Chief of Staff, and Lieutenant Colonel Bridges had been substituted for Lieutenant Colonel Barnardiston as British military *attaché* since 1906. The text of the document is as follows:

"The British military *attaché* asked to see General Jungbluth. The two gentlemen met on April 23.

"Lieutenant Colonel Bridges told the general that England had at her disposal an army which could be sent to the continent composed of six divisions of infantry and eight brigades of cavalry, together 160,000 troops. She has also everything which is necessary for her to defend her insular territory. Everything is ready.

"At the time of the recent events the British government would have immediately effected a disembarkation in Belgium, even if we had not asked for assistance.

"The general objected that for that our consent was necessary.

"The military *attaché* answered that he knew this but, that since we were not able to prevent the Germans from passing through our country, England would have landed her troops in Belgium under any circumstances.

"As for the place of landing, the military *attaché* did not make a precise statement; he said that the coast was rather long, but the general knows that Mr. Bridges during

Easter has paid daily visits to Zeebrugge from Ostende.

"The general added that we were, besides, perfectly able to prevent the Germans from passing through."

The German authorities and press have undoubetdly exaggerated the importance of these documents; but, on the other hand, they are not to be entirely ignored. They are merely the records of conversations between officers who had no authority to make any agreements binding their respective governments. The documents are not even signed by both parties like contracts or other formal conventions. But while possessing no intrinsic importance of themselves, they may have great historical value, if they reveal the intentions of the two governments.

We may quickly dispose of the first document. It appears from it that a British officer conferred with the Belgian Chief of Staff as to measures to be taken by Great Britain in case Belgian neutrality were violated by Germany. As one of the signatory powers of the treaty of 1839, Great Britain might justly regard herself as under obligation to intervene in such an event, and subsequent occurrences have proved that these precautions were not unwarranted.

But the later conversation, in the third document cited, contains indications of more serious appearance. Lieutenant Colonel Bridges asserted that at the time of some recent events, presumably the Moroccan crisis of 1911, the British government would have immediately effected a disembarkment in Belgium, even if the Belgian government had not asked for assistance.

A German commentary on the documents possessing official endorsement regards the above statement as absolute proof that the British government had the intention, in the case of a Franco-German war, to send troops to Belgium immediately, or in other words, to violate Belgian neutrality in anticipation of Germany's action. But even if we assume that Lieutenant Colonel Bridges was accurately informed as to the intention of his government, the sense of this passage depends upon the circumstances which are to be understood as the condition of the immediate intervention of the British forces. For these are not stated. Would the British disembarkment have immediately followed the declaration of war between Germany and France, or the violation of Belgium's neutrality by Germany? It is quite possible that Lieutenant Colonel Bridges meant to say that in case the Germans entered Belgium, Great Britain would immediately disembark her expeditionary forces in Belgium without awaiting Belgium's formal request for assistance. Let us assume, however, the most unfavorable construction for British intentions; let us suppose that Great Britain would have taken precisely the same step which Germany took in the present conflict, that she would have violated Belgian neutrality to anticipate her enemy. With such an interpretation the document would incriminate British intentions; but it would at the same time absolve Belgium by showing that General Jungbluth insisted that British intervention without Belgium's consent would be unlawful. This, and the statement that Belgium was quite able to prevent the Germans from passing through, might be interpreted as indications of an attitude in conformity with Baron Greindl's remonstrance. The Germans object that Belgium has never made any attempt to take, in agreement with the German government or military authorities, defensive measures against a possible Anglo-French violation of their territory. But it is fair to observe that the Anglo-Belgian conversations took place at the initiative of the British *attaché*, not the Belgian authorities. We may certainly assume that Germany never proposed any such exchange of views; for if she had, and her proposal had not been accepted, the circumstance would assuredly have been cited as evidence of Belgium's complicity by the German government or inspired press.

Our analysis of the documents leads us to regard their contents as insufficient to prove that the neutrality of Belgium as guaranteed by the powers was no longer valid in the summer of 1914. But to continue our argument to a further stage, we may ask ourselves what would have been the position of Belgium, if Germany's claim were established. The extreme view based on the German assumption regards Belgium as an ally of Great Britain, perhaps of France also. In that situation Belgium could no longer shield herself behind the barrier of the treaty of 1839. Nevertheless, until Belgium had committed some hostile act, or declared war, she would have been entitled by the law of nations to possess her territory inviolate like any other non-belligerent power. The alliance with a belligerent power is not in itself an adequate occasion for a declaration of hostilities. Germany did not declare war against France on account of the latter's alliance with Russia without alleging specific hostile acts committed by France.

It is true that the German authorities have alleged some breaches of neutrality on Belgium's part, such as the presence of French officers in Belgium, the flight of French air-craft and passage of French automobiles with French officers across Belgian territory. But scarcely any allusion is made to these occurrences at the time. They were neither mentioned as justification of their step in the public utterances of members of the German government, nor cited as motives in the course of diplomatic intercourse. At a later time these alleged violations of Belgian neutrality were published, and the statement was made that the German government possessed ample proof for their authenticity. The world has waited many months for a publication of this proof. But it is doubtful whether the alleged facts, even if proved, would constitute a sufficient cause for the German invasion of Belgium. For the present, German action in Belgium has no other justification than the motives expressed by her diplomatic chiefs at the time, considerations of strategic expediency.

Herr von Jagow replied to the remonstrances of the British ambassador, August 4, that the Germans "had to advance into France by the quickest and easiest way, so as to be able to get well ahead with their operations and endeavor to strike some decisive blow as early as possible. It was a matter of life and death for them, as, if they had gone by the more southern route they could not have hoped, in view of the paucity of the roads and the strength of the fortresses, to have got through without formidable opposition entailing great loss of time. This loss of time would have meant time gained by the Russians for bringing up their troops to the German frontier. Rapidity of action was the great German asset, while that of Russia was an inexhaustible supply of troops."

On the same day, in the course of his speech in the Reichstag, Chancellor von Bethmann-Hollweg expressed himself as follows:

"Gentlemen, we are now in a position of necessity; and necessity knows no law. Our troops have occupied Luxemburg; perhaps they have already entered Belgian territory. Gentlemen, this is in contradiction to the rules of international law. The French government has declared in Brussels that it is willing to respect the neutrality of Belgium so long as it is respected by the enemy. But we knew that France stood prepared for an inroad. France could wait, but we could not. A French inroad on our flank on the lower Rhine could have been fatal to us. So we were forced to set aside the just protests of the Luxemburg and Belgian governments. The wrong—I speak openly—the wrong that we now commit we will try to make good again as soon as our military ends have been reached. When one is threatened as we are, and all is at stake, he can only think of how he can hack his way out."

The Chancellor did not deny that Germany's action was morally indefensible. The German government acted on the view that the vital interests of a nation supersede its international obligations. But even on this last line of defense the German position regarding Belgium can scarcely be regarded as unassailable. Strategic considerations must be postponed for later discussion. But the opinion may be expressed in anticipation that the subsequent course of the campaign has not proved that the rapid drive through Belgium was the only salvation for German vital interests, even if the French would otherwise have advanced through that country to attack Germany.

But we are more directly concerned with historical causation than moral obligation. However justly the right of inviolability could be claimed for Belgium, the violation of Belgian neutrality was in only a limited sense the real cause of Great Britain's declaration of war, if the British government had practically engaged itself to support France before it had become known that Belgian neutrality was threatened.

It may be remarked that some German apologists have allowed themselves to be carried beyond the goal by their eagerness to attach moral turpitude to Great Britain's motives, so that their argument fails to produce the wished for effect. In their eagerness to impute a base motive to Great Britain, they deny her sincerity in going to war for Belgian neutrality, but to establish this point they proceed to prove that she had really been for some time bound by an engagement to help France. They do not recognize that there is no more disgrace in being an ally of France than an ally of Austria-Hungary. Their argument would leave Great Britain with an empty formal cause of war, it is true, but merely nominal alleged causes are accepted as an international conventionality without leaving a very serious moral stigma.

The supposed obligation to support France rests chiefly on the following facts:

The French fleet was concentrated mainly in the Mediterranean Sea, and the British fleet in the North Sea and adjoining waters. Trusting in the friendship of Great Britain, France left her northern coast unprotected by her own navy.

In 1912 there was the following exchange of views between Sir Edward Grey and M. Paul Cambon, the French Ambassador, relative to the possible military coöperation of the two nations:

Foreign Office, November 22, 1912.

My dear Ambassador,

From time to time in recent years the French and British naval and military experts have consulted together. It has always been understood that such consultation does not restrict the freedom of either government to decide at any future time whether or not to assist the other by armed force. We have agreed that consultation between experts is not, and ought not to be regarded as, an engagement that commits either government to action in a contingency that has not arisen and may never arise. The disposition, for instance, of the French and British fleets respectively at the present moment is not based upon an engagement to coöperate in war.

You have, however, pointed out that, if either government had grave reason to expect an unprovoked attack by a third power, it might become essential to know whether it could in that event depend upon the armed assistance of the other.

I agree that, if either government had grave reason to expect an unprovoked attack by a third power, or something that threatened the general peace, it should immediately discuss with the other whether both governments should act together to prevent aggression and to preserve peace, and, if so, what measures they would be prepared to take in common. If these measures involved action, the plans of the general staffs would at once be taken into consideration, and the governments would then decide what effect should be given to them. Yours, etc.,

E. Grey.

French Embassy, London,
November, 23, 1912.

Dear Sir Edward,

You reminded me in your letter of yesterday, 22nd November, that during the last few years the military and naval authorities of France and Great Britain had consulted with each other from time to time; that it had always been understood that these consultations should not restrict the liberty of either government to decide in the future whether they should lend each other the support of their armed forces; that, on either side, these consultations between experts were not and should not be considered as engagements binding our governments to take action in certain eventualities; that, however, I had remarked to you that, if one or other of the two governments had grave reasons to fear an unprovoked attack on the part of a third power, it would become essential to know whether it could count on the armed support of the other.

Your letter answers that point, and I am authorized to state that, in the event of one of our two governments having grave reasons to fear either an act of aggression from a third power, or some event threatening the general peace, that government would immediately examine with the other the question whether both governments should act together in order to prevent the act of aggression or preserve peace. If so, the two governments would deliberate as to the measures which they would be prepared to take in common; if those measures involved action, the two governments would take into immediate consideration the plans of their general staffs and would then decide as to the effect to be given to those plans.

Yours, etc.,

Paul Cambon.

Finally, when war was practically certain between France and Germany, the British government promised that the British fleet would protect the exposed French coasts against a hostile attack by the German fleet, and on August 2, Sir Edward Grey gave the following memorandum to M. Paul Cambon embodying the decision taken at a cabinet meeting that morning:

"I am authorized to give an assurance that, if the German fleet comes into the channel or through the North Sea to undertake hostile operations against French coasts or shipping, the British fleet will give all the protection in its power.

"This assurance is of course subject to the policy of His Majesty's government receiving the support of parliament, and must not be taken as binding His Majesty's government to take any action until the above contingency of action by the German fleet takes place."

In view of these facts, in spite of the formal statement that they did not strictly involve an obligation to bring mutual assistance, Great Britain would doubtless have exposed herself to accusations of disloyalty if she had not assisted France. But they do not constitute absolute proof that Great Britain would have gone to war if the Belgian cause had not arisen; while, on the other hand, British policy for two hundred years shows that Great Britain would not have stood idly by and permitted a great power to occupy any part of Belgium. For a vital British interest is involved in the independence and integrity of that country, upheld on many a field from Ramillies to Waterloo.

With a surer instinct for forcible presentation, Chancellor von Bethmann-Hollweg cites the promise of August 2 alone as a pledge for Great Britain's eventual coöperation with France, so that he lets the intervention of Great Britain appear as the result of a sudden impulse to profit by a favorable opportunity, not the natural consequence of a consistent policy. But the engagement of August 2 was clearly consistent with the policy that goes back to the redistribution of the naval forces of the two countries in such a way that their joint interests in the Channel were entrusted to the British fleet alone.

We may conclude, therefore, that the understanding with France was a very important factor. It is quite possible that Great Britain would have eventually joined forces with France, even if Belgian neutrality had not been violated. This understanding was not the result of an act of sudden stealth, but part of a respectable policy of several years' standing, the existence of which was generally recognized.

What were the ulterior causes of Great Britain's participation in the war? Only two, commercial rivalry and alarm at the growth of the German navy, are generally recognized. The Chancellor represents commercial jealousy as the prevailing, if not unique, cause for Great Britain's interference.

It is reasonable to assume that a war could arise from commercial jealousy only in case the prosperity of one country were confined, or seriously threatened, by the commercial activity of another nation. Statistics will readily show us whether such was the relation of affairs between Great Britain and Germany.

In the table on page 79 the per capita exports of the British nation display a very substantial growth from 1891 to 1911. Now national feelings are the complex product of

Prince Karl Max von Lichnowsky,
German Ambassador to Great Britain.

Jules Cambon,
French Ambassador to Germany.

Sir William Edward Goschen,
British Ambassador to Germany.

the feelings of individuals, and a nation of prosperous individuals is not likely to be impelled by dangerous sentiments of hostility on account of commercial jealousy. The figures for the aggregate exports of the United Kingdom at intervals of five years offer even more striking proof:

VALUE OF BRITISH EXPORTS

	Total exportation	Exportation to Germany
1891	$1,202,138,000	$91,389,000
1896	1,167,507,000	108,106,000
1901	1,360,908,000	114,568,000
1906	1,825,296,000	163,090,000
1911	2,209,972,000	190,920,000
1913	2,552,692,000	197,690,000

During the five years, 1891–1896, British exportation declined while German exportation at this time was increasing. According to the theory that commercial rivalry is a cause of hostility between Germany and Great Britain there should have been jealousy and international animosity at that time. But it is the only period of equal length since 1891 when the relations between the two countries were undisturbed. On the other hand, the situation became acute when both countries enjoyed unparalleled prosperity and the exports of Great Britain were increasing by leaps and bounds. Commercial rivalry may be relegated, therefore, to a remote place among the potential causes. If the two countries were going to be led into war as a direct consequence of commercial rivalry, the conflict would have occurred twenty years earlier, not in 1914.

There remains the apprehension of German naval development, and it is not necessary to repeat what has been shown with regard to it in Chapters III and IV. Alarm at the growth of the German navy had been the dominating factor in British foreign policy for a decade. The progress of German sea-power occupies a unique position among the potential causes of the war.

CHAPTER VIII

Was the War Deliberately Provoked?

The accusation against the Teutonic powers. Increase in Germany's military establishment. Extension of military railways on the Belgian frontier. Opening of enlarged Kaiser Wilhelm Canal, July 1, 1914. Russian military development. Internal conditions in Russia. Home Rule controversy in Ireland. The king's conference. Events in Dublin, July 26. Militarism; how it endangers peace. Expressions of the militaristic spirit; duelling, artificial conception of honor. Dangerous influence on political policies by insisting upon exaggerated standards of honor and prestige, and subordinating diplomacy to its professional point of view.

The circles which controlled the foreign policy of the Teutonic powers have been arraigned before the court of humanity on the charge of having deliberately provoked the present war. Obviously we cannot dismiss the subject of the causes of the war before we have examined the evidence for this indictment, which is brought by responsible individuals. For it cannot be denied that no genius in statecraft could have chosen a more propitious moment for so diabolical an undertaking, a moment when the nations of the Triple Entente were less prepared. There were alarming strikes in St. Petersburg a week before the commencement of the European war, which might at any time have ignited the inflammable material of widespread discontent in Russia and generated a revolutionary conflagration. President Poincaré and Prime Minister Viviani of the French Republic were absent from Paris on a visit to the Tsar, escorted by the two most effective units of the French fleet, which might have been bottled up in the Baltic Sea by a rapid German aggressive movement. As it was, the president and prime minister did not return to Paris until July 29, when the crisis was just entering upon its final, and most critical, stage. Several of the ambassadors of the *entente* powers in the most responsible positions were absent on leave from the capitals to which they had been accredited. The question regarding the attitude of the Ulster Protestants to Home Rule had taken on a most threatening aspect. Clearly these indications which obtrude themselves upon us are sufficiently striking to warrant a closer examination of the evidence.

For the sake of convenience we may divide the material into active and passive evidence, the first category including such positive acts of Germany or her allies as might appear to indicate a definite intention of engaging in hostilities; the second, the conditions prevailing in the *entente* powers or their dependencies, or elsewhere, which seemed to promise success for a military enterprise undertaken at this time. Clearly all possible evidence would be too voluminous for exhaustive treatment. In the first class particularly most of Germany's measures as related thus far in the present volume might be remotely involved. We must confine our attention to a few proceedings which may be brought into immediate and obvious connection with the war.

Allusion has already been made to the imposing augmentation of the German military establishment, which was well under way by the summer of 1914. Such an unprecedented enlargement is scarcely conceivable, in spite of the shifting of the balance of power in the Balkans, unless the leaders in Germany actually believed that a war would soon occur.

Early in 1914 the Germans completed a remarkable program of railway expansion in the direction of the Belgian frontier. Within a territory of 3600 square miles, which has the form of a triangle, lying between the frontiers of Belgium and Luxemburg and the Moselle and Rhine, the railway density was carried from 15.10 to 30.23 miles of line for

René Viviani, Prime Minister of the

Baron Wilhelm Edward von Schön, German Ambassador to

every hundred square miles of area during a period of five years. In particular, several new double-track lines were constructed from the Rhine to points of no intrinsic importance situated within a short march of the Belgian frontier. As this region is thinly populated and possesses neither important industries nor mineral deposits, such a sudden progress in railway development could have only a strategic significance.

On July 1, 1914, just five weeks before the outbreak of the war with Great Britain, the Kaiser proclaimed the formal opening of the enlarged canal which bears his name. The Kaiser Wilhelm Canal, extending from Kiel to the lower Elbe, and thus connecting the Baltic and North Seas, was first opened for traffic in 1894; but, with the enormous development in the size of ships, the dimensions of the canal became inadequate. Its commercial importance may be judged by the fact that ten times as many vessels pass through it annually as through the Suez Canal. In expanding its capacity its depth was increased from twenty-eight to forty-six feet, and its locks were enlarged to an extent that would permit them to accommodate the *Vaterland*, not to mention the largest war-ships in the German navy. The canal effects an economy of more than two hundred miles in the course of ships bound to the Baltic Sea from North Sea points lying southwards. From the naval point of view, the possession of such a canal is indispensable for Germany in any war involving operations by sea, at the same time, against Russia and one or both of the western powers. For it permits Germany to employ her entire naval force interchangeably in the Baltic and North Seas.

These facts, striking as they seem, may all be interpreted as defensive, precautionary measures. The German government always suspected that the real purpose of the Triple Entente was offensive. Thus in 1908, after the visit of

President Fallières at London, and the meeting of King Edward VII and the Tsar at Reval seemed to rivet the links of the *entente*, the Kaiser exclaimed at the German military maneuvers at Döberitz: "It really seems as if there is a desire to surround and provoke us. We shall know how to deal with it. The German has never known better how to fight than when he has had to defend himself on all sides. Let them come against us, we are ready!"

There is no doubt that Russia had been putting forth every effort since 1909 to augment the strength and efficiency of her military establishment. For instance the period of military service had recently been prolonged. It had been exactly three years, from January 1 until the termination of the third year. It was now prolonged until April 1 of the fourth year. Thus, if hostilities broke out early in the spring, the Russian government would have a very much stronger army ready at hand by merely retaining these most thoroughly trained fourth-year men under the colors. This measure was calculated to provide an active army of 1,840,000 men for the winter of 1913–14 and until April 1, 1914, of whom 1,322,000 would be serving in Europe, while there would be in all 1,415,000 during the remainder of 1914, with 1,017,000 serving in Europe. Moreover, Russia was developing her naval power with great rapidity. From 1909 until 1914 her annual expenses for naval purposes had increased 154%, as compared with 50% in France, 29.6% in Great Britain, and 13.8% in Germany.

The German government was undoubtedly aware that there was some kind of understanding between France and Great Britain relative to common action in case of hostilities. They also viewed with apprehension a possible naval convention between Great Britain and Russia. Even with the enlargement of the Kaiser Wilhelm Canal the presence of a considerable Russian naval force in the Baltic would be a constant source of uneasiness to Germany. The English opposition to the Dutch project of building a modern fortress at Flushing to command the entrance of the Scheldt was interpreted in Germany as an indication that Great Britain contemplated an offensive movement by way of Antwerp. The extension of the harbor of Esbjerg on the western coast of Denmark, eighteen miles from the German frontier, from 39 to 100 acres, with a very much larger extension in progress, was regarded as evidence of an aggressive design of Great Britain in that quarter.

In view of these circumstances, or, more especially, the feeling of alarm which they excited in Germany, the argument that the strategic railways and the enlargement of the canal, as well as the increase in the German army, were intended chiefly as defensive measures might appear plausible. Accordingly the evidence for Germany's culpability in provoking the war intentionally is thus far not convincing. We turn to the category of passive evidence.

Notwithstanding the undeniable material progress of Russia in recent years, signs and events were not lacking, during the months directly preceding the war, which might be regarded as indications of widespread dissatisfaction, as well as unpreparedness for any great national exertion or crisis. These phenomena were noticeable both in the borderlands and in the heart of the empire itself.

Allusion has already been made to the recent reappearance of the Russian nationalist tendency to undermine gradually the liberties of Finland. Stolypin, the forcible Russian Prime Minister, under whose administration the narrow spirit of nationalism had grown strong again, was assassinated in a theater in Kieff, September 18, 1911. The Finns had hoped that his successor Kokovtsoff would discontinue the hateful policy of Russification within the grand-duchy. But in this they were disappointed.

A law had been enacted in St. Petersburg, June 30, 1910, which opened a dangerous breach in the constitutional safeguards of Finnish liberty. It was virtually a reënactment of the odious imperial manifesto of 1899; for it reserved to the Tsar and Russian Duma the right to legislate on all Finnish questions which affected the interests of the empire as a whole. This far-reaching measure was never submitted to the Finnish Diet, and therefore the Finns refused to acknowledge its validity. But on the basis which it afforded, the Duma passed a law, February 2, 1912, establishing an equality of civil rights for Russians and Finns in the grand-duchy. It could be pointed out in justification of this law, that while 3,000,000 Finns enjoyed the full privileges of citizenship throughout the Russian Empire, the 150,000,000 Russians did not possess these rights in Finland. The Finns regarded the bill with anxiety, chiefly because it opened the civil service to Russians, permitting the substitution of Russians for Finns in all the important positions and the Russification of the administration.

Finnish officials refused to obey this law because it had not been sanctioned by the diet, and the courts upheld them in their defiant attitude. Accordingly, many officials and magistrates were taken to St. Petersburg to be tried for their contumacy; and, as a climax, the district court in the Tsar's capital tried and convicted the entire bench of the Viborg court of appeal, twenty-three judges, January 27, 1913. They were arrested, September 23, and conveyed to St. Petersburg in a special train to undergo their sentence of imprisonment, although in the meantime the Finnish Diet had passed laws assimilating the legal position of Russians to that of Finns in the grand-duchy. These judges were released after eight months of solitary confinement and received a tremendous ovation upon their return to Helsingfors. The Russian mounted gendarmes could scarcely disperse the excited crowds, who greeted with intense enthusiasm those who had made themselves national heroes by their violation of an unrighteous law. These occurrences made a deep impression and created a feeling of apprehension throughout the entire grand-duchy.

Of more immediate interest to the members of the opposite group of powers were some unmistakable signs of the recent progress of the national spirit in Poland. In the second Duma, when enthusiasm for the new constitutional régime was still fresh, an important group of Polish deputies under the leadership of Dmowski had expressed their willingness to cast their lot loyally with the Russian Slavs under the same constitutional government. But Poland had been slighted, together with the other border-lands represented in the Duma, in the great reduction of its representation embodied in the revision of the election laws, June 16, 1907. Furthermore, the third Duma sanctioned a bill, which became law, July 6, 1912, wounding Polish national pride unnecessarily by transferring the control of the district of Kholm from the Governor-General of Warsaw to the Minister of the Interior at St. Petersburg, practically violating the territorial integrity of the Kingdom of Poland, as it had been defined by the Congress of Vienna. The subsequent elections in Poland were clearly a setback for Dmowski's party of conciliation and sympathetic coöperation with Russia.

A feeling of discontent among the intelligent classes in Russia had been nourished by the government's continued application in many sections of such restrictive measures as are only justifiable in case of serious disturbances and disorder. All the parties in the Duma sharply criticised the government in 1913 for its reactionary methods, while at a meeting of the representatives of the various cities at Kieff resolutions in the same spirit were adopted.

We have observed that there is likely to be a dangerous connection between labor troubles and political unrest in Russia. Therefore, the statistics of strikes may be regarded as possible evidence of the popular attitude toward the established political order. The 1,051,000 strikers, as enumerated in 1905, fell to 142,000 in 1910, but rose again to 460,000 in 1913, which might readily be interpreted as a menacing portent.

In the early part of 1914, a letter from some German source in St. Petersburg was published in the *Cologne Gazette*, which emphasized the contrast between the existing defects and future possibilities of the Russian military resources. This article stated that Russia's vast gold reserve was scattered, largely lent out to smaller institutions throughout the country, so that it could not straightway be made liquid. Russia was at present incapable of effective mobilization on account of both military and financial weaknesses. The natural deduction seemed to be, that the present was precisely the time to undertake a preventive war against her.

The imminence of Home Rule in Ireland, and the determined spirit of resistance to it among the Protestant population in Ulster, created a very embarrassing and dangerous situation for the British government in the first half of 1914.

The sending of regular troops to strengthen the military depôts in Ireland precipitated a crisis in the British army, where the rumor quickly spread that these forces would be employed to coerce the people of Ulster into accepting Home Rule. A large number of officers resigned or demanded assurances from the government that their services would not be employed for the purpose of political coercion. Colonel Seely, Secretary of War, added to the Cabinet's reply to the resigning officers an assurance that the government did not intend to take advantage of its right to employ the military forces where it wished for crushing political opposition to its policy or the principles of Home Rule. But when Prime Minister Asquith repudiated Colonel Seely's statement, the latter resigned, and with him the Chief of Staff, Sir John French, who had supported Colonel Seely in his assurance to the officers. Prime Minister Asquith very unexpectedly assumed the portfolio of war.

This spirit of insubordination in the army, or as some indignantly called it, attempt at "military dictation to parliament," was doubtless regarded with much astonishment and interest in other countries. Late in April, 40,000 rifles and 50,000 rounds of ammunition, "made in Germany," were successfully landed on remote parts of the coast of Ulster in violation of the embargo which had been placed upon the importation of weapons by the government.

The government suggested as a concession for the purpose of a compromise and reconciliation, that the counties of Ulster should decide individually by ballot whether they would come immediately under the provisions of Home Rule, or preferred to be excluded from the application of the measure for a period of six years. But the opposition declared that this concession was not sufficient. The Home Rule Bill passed the House of Commons, as has been observed, May 25, 1914, the proposed conciliatory amendments being reserved for a later amending act, because it was necessary that the original bill should have been passed by the House of Commons three times in not less than two years, so as to overcome the veto of the House of Lords.

The situation in Ireland was rendered still more critical by the organization of bodies of Irish Nationalist volunteers. Thus Ireland was divided into two hostile camps, which seemed to be only awaiting the suitable occasion to commence hostilities. Moreover, an apparently hopeless deadlock prevailed in the negotiations for a compromise arrangement. The Unionists refused to accept the original bill, and Mr. John Redmond, leader of the Irish Nationalists rejected the conciliatory amending proposals as introduced in the House of Lords.

Finally, King George took the initiative in a new conciliatory movement by calling the leaders of both sides to a conference in his presence. The first session of this conference occurred on July 21, and it is most interesting, during that time, when each day was teeming with significance for the nations, to coördinate these proceedings in Great Britain with the startling events on the continent.

The personal initiative of the king was unfavorably regarded by many people on principle; and the king's somewhat tactless allusion to the "cry of civil war" as being on the lips of the people seemed to be among the causes that foredoomed this conference to failure. It separated without accomplishing anything.

The occurrences on Sunday, July 26, near Dublin, doubtless appeared to German statesmen as the commencement of the predicted civil war which would fatally embarrass Great Britain's action in world-affairs. The sanginuary encounter grew out of an attempt to land a consignment of 10,000 rifles near Dublin for the Nationalist volunteers. Troops were sent to seize these weapons, and upon their return march towards Dublin the soldiers were attacked by a mob and pelted with improvised missiles. When their patience became exhausted, they fired into the rabble, killing four persons and wounding about sixty.

Thus a strong case may be made out against Germany, although it rests entirely upon circumstantial evidence. On the other hand, against the charges, the argument may be offered that if the Teutonic powers had intended to seize a favorable occasion for a deliberate attack upon the rival powers, they would have done this before the military power of Turkey, or that of Bulgaria, had been crushed in

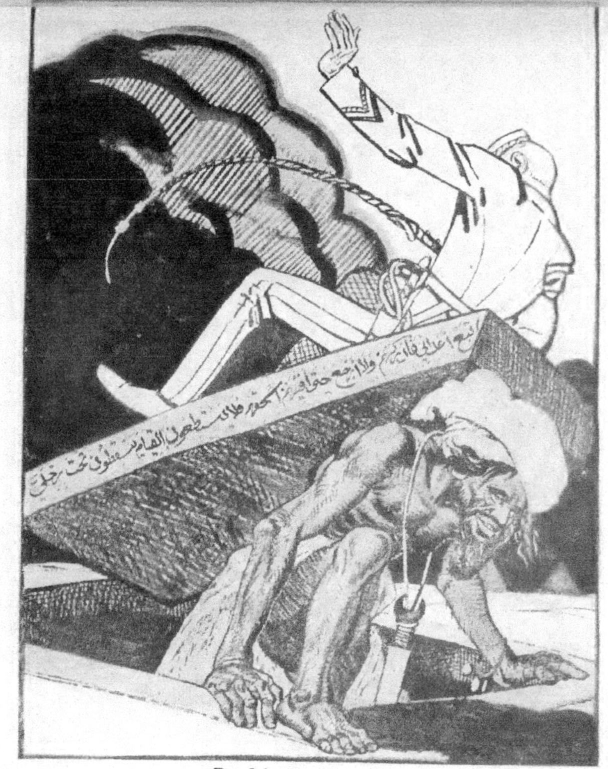

Der Islam rührt sich! Julius Diez (München)

Islam bestirs himself! Cartoon, which appeared in *Jugend*, showing German expectation that Great Britain would be embarrassed by a Mohammedan revolt.

„Goddam! Jetzt wird mir meine Kundschaft nicht mehr glauben, daß die deutschen Stahlwaren schlecht sind!"

England's Pain. Cartoon, which appeared in *Simplicissimus*, showing German feeling in regard to the commercial aspect of the war.

the Balkan conflicts. The fact that Germany and Austria-Hungary did not employ favorable opportunities in 1912 or 1913 to attain by force the aims which are ascribed to them in the present war is evidence that they did not contemplate an aggressive action at that time. This argument may very reasonably have the effect of narrowing the charge. It was not part of the general policy of the Teutonic powers to overwhelm the rival group by an unexpected assault at the first suitable opportunity. In consequence of this, unless we are prepared to accept the sensational theory that the murder of the Archduke was part of the larger Teutonic plot for sudden aggressive action, intended to provide an appropriate occasion, we are not prepared to admit the existence before June 28, 1914, of an Austro-German plan for deliberately provoking hostilities.

In judging the motives which came into play after the murder of the Archduke Francis Ferdinand, it is necessary to observe that there are strong indications in support of the opinion that the Teutonic powers believed that they would realize their purpose regarding Serbia, and whatever that implied in connection with influence and prestige in the Balkan peninsula, without precipitating a general European war; in other words, that in consequence of the weaknesses that have been enumerated, the *entente* powers were not merely doomed to succumb in case of actual hostilities, but would be unwilling to fight.

The following extracts from a report by Sir M. de Bunsen, British Ambassador at Vienna, of a conversation with his German colleague, July 26, may be cited as evidence for this attitude:

"According to confident belief of German Ambassador, Russia will keep quiet during chastisement of Serbia, which Austria-Hungary is resolved to inflict, having received assurances that no Serbian territory will be annexed by

Austria-Hungary.". . . . ."The Russian Minister for Foreign Affairs would not, his Excellency thought, be so imprudent as to take a step which would probably result in many frontier questions in which Russia is interested, such as Swedish, Polish, Ruthene, Roumanian, and Persian questions, being brought into the melting-pot. France, too, was not at all in a condition for facing a war."

Likewise, on July 28, the Austrian ambassador in Berlin informed Sir Edward Goschen "that a general war was most unlikely, as Russia neither wanted nor was in a position to make war." The British ambassador believed that the same opinion was shared by many people in Berlin. Sir R. Rodd, British Ambassador in Rome, made a report on the 29th in the same sense, stating the opinion of the Italian Foreign Minister "that there seemed to be a difficulty in making Germany believe that Russia was in earnest." The Russian Foreign Minister Sazonoff expressed the conviction that Germany was unfortunate in her representatives in Vienna and St. Petersburg, since "the former was a violent Russophobe who had urged Austria on," while "the latter had reported to his government that Russia would never go to war."

Again, Great Britain is represented as the reluctant party in the *Triple Entente*, whose attitude of weakness nourished the confidence of the Teutonic powers. The Russian ambassador told Sir Edward Grey on the 27th that the impression prevailed in German and Austrian circles that in any event Great Britain would stand aside.

If the theory is adopted that the Germanic powers believed that they could win a bloodless, diplomatic victory by maintaining an attitude of firmness, this conviction supersedes, for the practical purposes of history in dealing with the causes of the war, the problem of the existence of an intention of provoking a general conflict.

We ought to note in this connection the evidence indicating a belief on the part of representatives of the *entente* powers that they could restrain their opponents, and thereby prevent the outbreak of hostilities, by adopting a steadfast, resolute bearing.

At the outbreak of the crisis the Russian Foreign Minister expressed the belief that Germany did not really want war but that her attitude was decided by that of Great Britain. For, as he declared, if Great Britain took her stand firmly with France and Russia, there would be no war; but if Great Britain failed them, rivers of blood would flow, and she would eventually be dragged into the war.

Sir F. Bertie, British Ambassador in Paris, reported to Sir Edward Grey on the 30th that the President of the Republic was convinced that peace between the powers was in the hands of Great Britain; that "if His Majesty's government announced that England would come to the help of France in the event of a conflict between France and Germany,..........there would be no war."

In a letter directed to King George, July 31, urging the advantage of Great Britain proclaiming herself openly the ally of France and Russia, President Poincaré declared his belief that "if Germany were convinced that the *Entente Cordiale* would be affirmed in case of need, even to the extent of taking the field side by side, there would be the greatest chance that peace would remain unbroken."

On the same day, M. Cambon in his representations to Sir Edward Grey, pointed out that it was the uncertainty as to whether Great Britain would intervene which was the encouraging element in Berlin, and that, if Great Britain "would only declare definitely on the side of Russia and France, it would decide the German attitude in favor of peace."

Thus it appears that the conviction was entertained in

influential circles on both sides that an exhibition of firmness was alone necessary for guaranteeing success.

But the problem of motives is complicated by the presence in most or all of the belligerent nations of a factor which we have not thus far directly considered—its mention requires a definition—militarism.

Some writers protest that the term militarism is very often thoughtlessly or incorrectly employed, and in particular, that it is erroneous to impute this quality with its odious implication to Germany; because Russia has a very much more numerous army, and Great Britain spends much more annually on her military and naval establishments. In fact, German apologists, in self-defense, have coined the expression navalism, which they impute to Great Britain with the same unfavorable significance.

But the true meaning of the term militarism has no necessary connection with the size of armies or magnitude of military establishments. It denotes a certain attitude of mind. The suffix *ism* very frequently implies an exaggerated or partisan view or support of the quality, attitude, or program of conduct indicated by the word to which it is attached. Thus militarism may be defined as an arrogant, or exclusive, professional military spirit. It may make itself noticeable in small military establishments, and yet is not necessarily conspicuous in large ones. Like the professional spirit in any other occupation, militarism is likely to exhibit extravagant examples in any country.

The significance of militarism will become clearer as we consider the dangers which it involves. There are two ways in which militarism may imperil peace. In the first place, it encourages both in individuals and states an exaggerated sensitiveness, or touchiness, and an artificial conception of honor.

On August 3, 1905, a so-called Court of Honor at Mülheim,

in Germany, dismissed ignominiously from his commission, as Landwehr Lieutenant, Dr. Fritz Feldhaus, because he did not challenge an officer who had insulted him, notwithstanding the fact that duels are legally forbidden, and Feldhaus had already obtained ample satisfaction by regular judicial proceedings. The matter was the subject of an indignant interpellation in the Reichstag the following January. The Prussian Minister of War observed, in replying to the interpellation, that although the duel as a means of settlement for questions of honor was still recognized extensively among the better classes, an imperial decree of 1897 had very effectively curtailed the practice among officers, and that the prospective revision of the criminal code with the introduction of severer penalties against insults would reduce it still further. "But as long as the duel is recognized in extensive circles as a means for healing injured honor, the corps of officers cannot tolerate in their midst any member who is not ready to defend his honor sword in hand."

These words provoked expressions of indignation from all parties except the Conservatives. It appeared from the minister's remarks that militarism has its own exclusive code of morals; and that in upholding an artificial conception of honor it disregards law and the common consciousness of right.

The practice of duelling among votaries of an unreasonable standard of honor is harmless as compared with the tyranny of imposing their bigoted standard upon others, and subjecting reason and conscience to its haughty requirements. Their despotic doctrine assumes superiority over rational motives and deliberations. It insists that under certain circumstances states as well as individuals must commit themselves to bloodshed and destruction without regard to considerations of right and wrong. It enslaves reason and justice alike to its exaggerated conception of honor. It jealously guards the sanctity of this holy of holies from the contamination of peace conferences, arbitration, and all other vulgar devices for the elimination of conflicts.

We considered at some length the incidents of the celebrated Dreyfus case, in consequence of which the militarism that had existed in a dangerous form in France received a merited rebuke. A military tribunal decided in advance that Captain Dreyfus must be condemned, whether guilty or not, to maintain military prestige, or satisfy military "honor." Such "honor" is as dishonorable as anything in this world. It is the foe of truth, reason, and morality.

Another flagrant exhibition of insolent militarism has a special interest as illustrating, in a one-sided way perhaps, the popular attitude in Alsace. A feeling of irritation existed for some time between the townspeople of Zabern and the soldiers of the regiment which was stationed in garrison in their midst. The report spread through the town that a young officer, Lieutenant Forstner, had announced to the recruits in the barracks that anybody stabbing a *Wackes* who insulted him would receive a present of ten marks. The term *wackes*, meaning literally toad in the local dialect, is employed by the Alsatians in addressing one another as a term of familiarity or endearment, but they resent this designation as an affront when it is applied to them by outsiders. Indignation was naturally aroused at Lieutenant Forstner's words, and, when some officers were jeered at on the street in the evening, Colonel Reuter ordered sixty soldiers with loaded rifles to arrest all civilians who did not retire. Twenty-seven persons were arrested in this summary fashion, including a local magistrate and a member of the fire department who had rushed into the street upon hearing the bugle. All these were confined over night in the cellars of the barracks. Besides this, Lieutenant

Forstner wounded with a sabre thrust a lame cobbler, who as he claimed, had insulted him, although this was denied.

This matter was the subject of an interpellation in the Reichstag, December 3.

The Chancellor, although not denying that the action of the military authorities at Zabern had been unjustifiable, endeavored to make the affair seem unimportant. He announced that Lieutenant Forstner and one non-commissioned officer would be punished. There was a storm of indignation as the Reichstag contemplated this violent intrusion of the military into the sphere of civil life, and the Chancellor seemed to have failed entirely to treat the really serious features of the situation. A resolution of disapproval of the Chancellor's treatment of the affair passed the chamber by a vote of 293 to 54. This is the second time that a vote of censure against a Chancellor has passed the Reichstag. Prince von Bülow had received a vote of disapproval for the government's Polish policy. In a state with a responsible government the cabinet would have resigned immediately. Chancellor von Bethmann-Hollweg did not resign. Lieutenant Forstner was condemned to forty-three days' imprisonment; but Dr. von Jagow, the police president at Berlin, deplored Forstner's punishment in an article in the *North German Gazette*, in which he said that "military exercises are acts of sovereignty, and, if obstacles are placed in the way of their performance, the obstacles must be removed in the execution of the acts of sovereignty."

Perhaps the words of the Kaiser in dedicating a monument to Prince Frederick Charles at Frankfurt-on-the-Oder in 1891 may be cited to illustrate an exaggerated, unnatural conception of military honor propagated by the militaristic tradition. He declared:

"We would rather sacrifice our eighteen army corps and our 42,000,000 inhabitants on the field of battle than surrender a single stone of what my father and Prince Frederick Charles gained."

After making a liberal discount for rhetorical exaggeration, the passage must still be interpreted in the sense that lives must be lavishly sacrificed, in certain situations, in vindication of an empty principle of military honor, without any useful purpose. If warfare can alone be justified as the final means for deciding important issues, how can the expenditure of lives and treasure be justified from the moment when events have made the result of the conflict certain?

The thrill evoked by such an address is an inherited, traditional kind of thrill, the product of an inveterate emotional habit. To the spirit of the present age, if it were free from historical bias, the Kaiser's heroic resolution would suggest the notion of suicide in consequence of a business liquidation where the assets turned out to be greatly in excess of the liabilities. Militarism flourishes in the daylight of modern ingenuity and efficiency, but it draws emotional sustenance from roots cast deep into bygone ages.

Frequent references to Austrian "prestige" in the diplomatic correspondence suggest that the spirit of militarism exercised an important influence upon the attitude of the government in Vienna.

Thus Sir R. Rodd reported as early as July 23, before the terms of the Austrian note to Serbia had been published, that the Italian Foreign Office expressed the view "that the gravity of the situation lay in the conviction of the Austro-Hungarian government that it was absolutely necessary for their *prestige*, after the many disillusions which the turn of events in the Balkans has occasioned, to score a definite success."

On July 28 the Austro-Hungarian Minister of Foreign Affairs informed Sir M. de Bunsen that Austria-Hungary could not delay military operations against Serbia any

Count Shigenobu Okuma, Prime Minister
and Minister of Interior, Japan.

Yoshihito,
Emperor of Japan.

Baron Taka-Akira Kato,
Minister of Foreign Affairs of Japan.

longer. "The *prestige* of the Dual Monarchy was engaged, and nothing could now prevent conflict."

Professor Munroe Smith, in discussing the problem of the relations of military strategy and diplomacy with particular reference to the events leading up to the great war, points to evidence of the existence at Vienna of a "touchiness closely related to the soldier's and duellist's sense of honor." Thus Herr von Jagow reminded Sir E. Goschen, July 29, that "he had to be very careful in giving advice to Austria, "as any idea that they were being pressed would be likely to cause them to precipitate matters and present a *fait accompli*." He was not sure that his communication of Sir Edward Grey's "suggestion that Serbia's reply offered a basis for discussion had not hastened the declaration of war" by Austria-Hungary. On July 30 the Chancellor told Sir E. Goschen "that he was not sure whether he had not gone so far in urging moderation at Vienna that matters had been precipitated rather than otherwise."

Militarism is dangerous, in the second place, because it is inclined to regard all issues solely from the professional point of view. This danger has increased with the more rigid specialization in the military profession coincident with the enormous development in the efficiency and accuracy of adjustment of all the parts of the military machine, and in the capacity and necessity for rapidity of action. The existence alone of a highly organized military machine is a menace to peace, unless it is under the control of persons who take more than a merely mechanical view of the purpose and function of this formidable apparatus.

To satisfy professional requirements militarism strives to impose its will upon the political leadership of the nation. It is ready to sacrifice the moral and political advantages of a defensive attitude or of deferring hostilities to the strategic advantage of the sudden attack.

The influence of militarism was a dominant factor toward the close of the period of discussion which led to the outbreak of hostilities.

Thus on July 30, M. Jules Cambon, French Ambassador at Berlin, reported to his government: "The military authorities urged strongly that mobilization should be decreed on the ground that any delay will lose Germany some of her advantages. Up to the present, however, it has been possible to resist successfully the haste of the General Staff, which in mobilization sees war."

On August 1, Herr von Jagow explained Germany's declaration of war against Russia on purely strategic grounds, as reported by Sir E. Goschen:

"Russia had said that her mobilization did not imply war, and that she could perfectly well remain mobilized for months without making war. This was not the case with Germany. She had the speed and Russia had the numbers and the safety of the German Empire forbade that Germany should allow Russia time to bring up masses of troops from all parts of her wide dominions."

Finally, when Great Britain's ultimatum was presented, Herr von Jagow explained "that they had to advance into France by the quickest and easiest way, so as to be able to get well ahead with their operations and endeavor to strike some decisive blow as early as possible. It was a matter of life and death for them, as, if they had gone by the more southern route they could not have hoped, in view of the paucity of roads and the strength of the fortresses, to have got through without formidable opposition entailing great loss of time. This loss of time would have meant time gained by the Russians for bringing up their troops to the German frontier. Rapidity of action was the great German asset, while that of Russia was an inexhaustible supply of troops."

Whatever judgment we form concerning the policy of the Russian authorities in taking the decisive step of ordering their general mobilization in advance of Austria-Hungary and Germany, we can scarcely absolve them from grave responsibility. The various aspects of the case will be discussed later; it suffices for the present purpose to mention the evident existence of disagreement in the higher counsels of the Russian imperial government. The Tsar was undoubtedly very much opposed to war; and in his desire to maintain a conciliatory attitude, he was supported by the diplomatic department of the government, and probably by the majority of the cabinet. But this policy was vigorously opposed by a military clique, who doubtless regarded Russia's military "honor" as at stake since 1909, and who probably insisted, from the professional point of view, that a further delay in the general mobilization would be courting disaster.

CHAPTER IX

Motives of Japan, Turkey and Italy

Transformation of Japan. War with China, 1894–5. Interference of Germany, France, and Russia. Japan deprived of her spoil. Russia in Manchuria and at Port Arthur. Anglo-Japanese Treaty signed, January 30, 1902. Russo-Japanese War; its termination by the Treaty of Portsmouth, September 5, 1905. Second Anglo-Japanese Treaty. Russo-Japanese understanding. Japan's present ministry. Japan's ultimatum and declaration of war against Germany. Position of Turkey at the outbreak of the war. Aspirations of Pan-Islamism. The Turkish ministry, Said Halim Pasha, Djavid Bey, Djemal Pasha, Talaat Pasha, Enver Pasha. German military commission. German influence. British intentions misrepresented. Arrival of *Goeben* and *Breslau*, August 10. Turkish duplicity. Capitulations abolished. Turkish military preparations. British protests. Commencement of hostilities. Jihad, or holy war, proclaimed. Motives for Italian hostility for Austria, sentimental, strategic, economic. Italy's domestic politics. Declaration of neutrality. Albanian occurrences. Austro-Italian negotiations; three stages; the intervention of Prince von Bülow, December 20, 1914; Austria's definite offer, March 27, 1915; Italy's counter-proposal, April 8; the repudiation of the Triple Alliance, May 4. The triumph of the Salandra ministry; the session of the Chamber, May 20; the declaration of war, May 23.

The complete transformation in the national life of Japan which took place during the second half of the nineteenth century constitutes the most unique chapter in human records, one for which an analogy is sought in vain. The story of this remarkable transition has become a classic. In the seventeenth century Japan shut herself off from intercourse with the outside world. Her institutions became rigid, her national life stationary. She retained until the middle of the nineteenth century a state of feudalism and methods of life almost as antiquated as those of mediæval Europe, and even the inclination for progress was absolutely non-existent. The nation was jealously protected against any contaminating suggestion of change by the stringent legal exclusion of foreigners.

The Mikado, or emperor, dwelt in mysterious seclusion in a palace at Kioto. Although regarded with a half-religious reverence, he possessed only the shadow and title of sovereignty. For the Shoguns, who were originally ministers, obedient to the Mikado, had gradually appropriated all the actual prerogatives of imperial authority. The Shogunate had become hereditary. The Shogun held court in Yedo, and stood at the head of the military aristocracy, the ruling class. The arrival of Commodore Perry in command of an American squadron in Japanese waters in 1853 broke the enervating spell that had brooded over Japan for so long a time. In 1854 two Japanese ports were opened to American shipping and commerce, and similar arrangements were very soon effected between Japan and the other leading commercial nations.

A fermentation in the national spirit was set in action, and it rapidly increased in intensity. The succession of Mutsuhito as Mikado, in 1867, was the signal for the culmination of the revolution. He resolutely stepped forth from retirement and displayed himself to the eyes of his subjects. He resumed the absolute authority of his forefathers ages back, abolished the Shogunate, and removed his residence from Kioto, the old capital, to Yedo, which was afterwards called Tokio, in 1868. Within a very few years western civilization was introduced into Japan in the most systematic manner.

The Japanese nation, whose civilization from the material point of view was comparable with that of Europe five hundred years ago, and whose life and institutions were entirely alien to those of the western nations, set out deliberately to recast its entire social, political, and industrial organization to accord with the most approved European standards of the present age. History offers no parallel for such a universal acceptance by a whole people of a policy of transformation affecting profoundly its most essential habits of life based solely upon an intellectual, unemotional conviction of expediency. It probably means that the Japanese excel all the other peoples in the world in the quality of national self-discipline.

The nineteenth century has recorded many instances of national transformations. But of all, the Japanese was at the same time the most sudden and comprehensive and the most unexpected. It violates all the rules of historical likelihood. For, while the transformation of other nations has been mainly nothing more than an acceleration of progress along national, or at any rate, racial lines, the Japanese have adopted with their marvellous capacity for assimilation a civilization which was the very antipodes of their own. All the customary institutions of a first class modern state were quickly introduced, newspapers, national schools and a university, scientific codes of laws, railways, universal obligatory military service—finally, a constitution in 1890. Within thirty years this wonderfully endowed people had covered the centuries which separated them from the great western powers, and were ready to enter into competition with them.

As Japan had an area of only about 140,000 square miles, which was so mountainous that only one-sixth was available for cultivation, her population of more than 40,000,000 was rather crowded in their island home. Japan required an opportunity for territorial expansion.

A question of suzerainty over Corea involved Japan in a war with China in the autumn of 1894; and as the vast Chinese Empire was still in the backward condition from which Japan had only begun to emerge about thirty years earlier, the result of the conflict could not long remain doubtful. Japan suffered the loss of scarcely more than five hundred soldiers killed in action in reducing an empire of about 400,000,000 souls to sue for terms of peace. The exaggerations of romance must yield the precedence to reality when compared with the facility with which this important enterprise was performed. By the Treaty of Shimonoseki, April 17, 1895, China recognized the independence of Corea, and ceded Port Arthur with the Liaotung peninsula, Formosa, and the Pescadores Islands to Japan.

At this time Germany, France, and Russia began to display an unusual solicitude for the welfare of the peoples of the Far East, like some officious individuals who are apparently more concerned about the weal of others than those others are about themselves. Russia as the spokesman of this trio represented to Japan in a quite friendly spirit that Port Arthur was the key to Peking and to the heart of China, and that its occupation by the Japanese would be a perpetual menace to permanent peace in the Far East. In consequence of this diplomatic intervention of the three powers mentioned, Japan was excluded from a foothold on the continent of Asia. Japan always believed that Germany was the instigator of this action of solicitude for the "permanent peace in the Far East," just as she was thought to have encouraged Russia in a campaign of expansion and adventure in that direction. If this suspicion is true, one might say that just as mankind is condemned before birth by an act of original sin, so the doom of Kiau-Chau was predetermined by a fatal act of transgression before this German colony came into existence. For Japan waited nineteen years to satisfy her resentment against Germany for the disappointment inflicted upon her in 1895.

It was not long before the western powers, which had interceded in China's behalf, began to assume rewards for their magnanimity. Russia was then engaged pushing forward her Trans-Siberian Railway across Asia towards

Vladivostok, a port on the Sea of Japan. She obtained a concession from China for building the main line across Northern Manchuria, and also for the construction of a branch line from Harbin to Talienwan in the Liao-tung peninsula, an ice-free harbor. This concession included the privilege of guarding these lines on Chinese territory with Russian soldiers, under cover of which Russia threw large forces into Manchuria. On March 5, 1898, Germany concluded a convention with China for a ninety-nine year lease of Kiau-Chau and adjacent territory, which virtually brought the Province of Shantung under her sphere of influence. Just about three weeks later, Russia concluded a lease of the district of Kwangto for a period of twenty-five years, and began to fortify the important position, Port Arthur, included in her leasehold, which, as she had declared to Japan, commands the approach to Peking and the heart of China. About the same time, France and Great Britain obtained similar leaseholds for themselves on the coast of China. The speedy disintegration of the Chinese Empire was predicted, and it is quite likely that this expectation was one of the chief motives for the greater naval activity of Germany which began at just this time.

But after the Boxer movement in China in 1900, and the international expedition to relieve the besieged legations in Peking, the powers guaranteed the integrity of the Chinese Empire.

The occupation of Southern Manchuria by Russia and the impending penetration of Corea by the same power not only threatened to exclude Japan from her normal field of expansion, but seemed even to menace her safety at home. The Japanese put forth every exertion to prepare for the impending struggle with the mighty colossus of the north.

Russia promised to withdraw from Manchuria when order was established, but that might very likely be a mere sub-

Enver Bey, Turkish Minister of War.

Djemal Pasha, Turkish Minister of Marine.

terfuge for continued occupation; and the conduct of Russia in the Far East was not calculated to inspire confidence in the sincerity of her protestations. At this time Great Britain and Japan became convinced of their common interest in restricting the aggrandizement of Russia and maintaining the *status quo* in the Far East.

A convention embodying terms of an Anglo-Japanese defensive alliance was signed in London, January 30, 1902. The treaty set forth the desire of the two governments to maintain the general peace in the East and the independence and territorial integrity of China and Corea, and to secure equal opportunities in those countries for the commerce and industry of all nations. It contained, also, a mutual recognition of the special interests of the two contracting parties.

The reciprocal obligations of the contracting parties were defined in the second and third sections as follows:

2. "If either Great Britain or Japan, in the defense of their respective interests as above described, should become involved in war with another power, the other high contracting party will maintain a strict neutrality, and use its efforts to prevent other powers from joining in hostilities against its ally.

3. "If in the above event any other power or powers should join in hostility against that ally, the other high contracting party will come to its assistance and will conduct the war in common and make peace in mutual agreement with it."

An alliance on an equal footing with one of the leading nations of Europe was naturally a source of pride to the Japanese. It likewise gave them greater assurance by eliminating the likelihood of a hostile coalition of powers against them in case of war with Russia.

Negotiations relative to the occupation of Manchuria

were carried on between Japan and Russia from August, 1903, until February, 1904, when the Japanese, convinced that it was futile to continue them any longer, made a sudden attack without any declaration of hostilities, torpedoing several war vessels in the harbor of Port Arthur during the night of the 8th and 9th. In August, 1904, the Japanese, in separate naval engagements, crushed the Russian fleets which had been stationed at Port Arthur and Vladivostok respectively, and on January 1, 1905, General Stoessel surrendered Port Arthur to General Nogi. Allusion has already been made to the startling effect of this disaster upon the temper of the Russian people. A sanguinary battle near Mukden continuing for several days resulted in a momentous victory for the Japanese. Finally, the Russian fleet which had been sent out from the Baltic, was shattered by the Japanese during an engagement in Tsushima Straits, May 27, 1905. The Treaty of Portsmouth, signed September 5th, recognized Japan's paramount interests in Corea. Both Russians and Japanese pledged themselves to evacuate Manchuria. Russia ceded to Japan the southern half of the island of Saghalin, and her lease of Kwangto, the southern extremity of the Liao-tung peninsula, which includes the seaports of Port Arthur, Dalny, and Talienwan. The possession of Kwangto includes also the control of the Southern Manchurian Railway.

A new treaty of alliance between Japan and Great Britain was signed, August 12, and published, September 27, 1905. It described the objects of the alliance as the defense of the "open door" in China, the maintenance of peace in Eastern Asia, and the protection of the special interests of the two contracting parties. The second article defined the reciprocal obligations of the contracting parties as follows:

"If by reason of unprovoked attack or aggressive action wherever arising, on the part of any other power or powers, either contracting party should be involved in war in defense of its territorial rights or special interests mentioned in the preamble of this agreement, the other contracting party will at once come to the assistance of its ally, and will conduct the war in common, and make peace in mutual agreement with it."

By 1907 Japan and Russia had arrived at complete harmony of views respecting the Far East, and on June 30 of that year a convention between them was signed at St. Petersburg, by which each power pledged itself to respect the territorial integrity and treaty rights of the other, expressed its formal recognition of the independence and territorial integrity of the Chinese Empire "as well as the principle of equal opportunity in commerce and industry for all nations in the said empire."

The volume of the economic life of Japan is still modest as compared with that of the most prominent European nations. Her area amounts to 140,191 square miles, or, including the dependencies, 235,886; her population was 50,052,798 in 1908, and including her dependencies, 67,142,798. Her annual budget balances at about the equivalent of $300,000,000 in American currency. Her total foreign trade represents an annual value of about $550,000,000. There were 5606 miles of railway lines in Japan proper in 1913. The present Mikado, Yoshihito, who was born August 31, 1879, and succeeded Mutsuhito at the latter's death, July 30, 1912, is an able and progressive ruler, and he is surrounded by a group of statesmen of keen intelligence and ample experience.

But the records of Japan's internal affairs during quite recent years do not display the same harmony and unanimity which have characterized the Japanese people in the most critical periods. This is due to the struggle of the popular parties against bureaucracy and militarism and in favor of complete parliamentary control of the government. The prime minister does not owe his appointment to the prevailing party in the popular house of the legislature; he is chosen by the Council of Elder Statesmen. There was a cabinet crisis at the end of 1912, when the question of popular influence was involved. The people demanded a reduction of the heavy taxation which had prevailed since the war with Russia. The Prime Minister, Saionji, was favorable to a policy of retrenchment, but the minister of war refused to consent to the proposed economy in military expenditure. The bureaucratic influence was rendered very emphatic by the customary reservation of the portfolios of war and marine to representatives of distinct clans. Saionji resigned, and Prince Katsura formed a ministry, which lived only until February, when it was overthrown by parliament. Then Admiral Yamamato was entrusted with the formation of a new cabinet. Popular dissatisfaction accompanied by reported scandals in the government continued, and the Yamamato ministry resigned March 25, 1914. The cabinet was then formed which is still in existence. Count Shigenobu Okuma is prime minister, one of Japan's foremost statesmen, and so well known throughout the world, that he is commonly regarded by foreign nations as the leading authority on the affairs of his own country. He was born in 1838, and was one of the foremost advocates of the abolition of the former superannuated institutions and the introduction of constitutional government. He organized the Progressive party, the forerunner of the present National party. He has been finance minister altogether during eleven years, and was premier for a short period in 1898, after which he retired for a time from active party leadership, founded Waseda University and became its president. Baron Taka-akira Kato holds the portfolio of foreign affairs in the Okuma ministry. He was born in 1859, served in London as minister plenipotentiary,

1894–99, and again as ambassador, 1908–12, and had already been minister of foreign affairs three times, the last time in the short-lived Katsura cabinet, before he was summoned to the same post in the present ministry.

The Japanese had always felt a sense of annoyance at the presence of the Germans in Kiau-Chau, similar, no doubt, though less intense than their resentment at the Russian occupation of Port Arthur. They welcomed the opportunity afforded by the present war to rid themselves of unwelcome neighbors.

According to the Anglo-Japanese treaty of 1905, which was still in force, if either contracting party should be involved in war in defense of its interests in the Far East in consequence of an unprovoked aggressive action of another power, the other contracting party should come to the aid of the first. It required a somewhat broad interpretation of the circumstances, perhaps, to construe the presence of the Germans at Kiau-Chau as a *casus belli* under the terms of the treaty of alliance with Great Britain. Still, it was conceivable that the presence of this hostile base injured British interests. After previous consultation with Great Britain, on Monday, August 16, 1914, the Japanese government directed its ultimatum to the German government in the following terms:

"We consider it highly important and necessary in the present situation to take measures to remove the causes of all disturbance of peace in the Far East, and to safeguard general interests as contemplated in the Agreement of Alliance between Japan and Great Britain."

"In order to secure firm and enduring peace in Eastern Asia, the establishment of which is the aim of the said agreement, the Imperial Japanese government sincerely believes it to be its duty to give advice to the Imperial German government to carry out the following propositions:

1. "To withdraw immediately from Japanese and Chinese waters the German men-of-war and armed vessels of all kinds, and to disarm at once those which cannot be withdrawn."

2. "To deliver on a date not later than September 15th to the Imperial Japanese authorities, without condition or compensation, the entire leased territory of Kiau-Chau, with a view to the eventual restoration of the same to China."

"The Imperial Japanese government announces at the same time that in the event of its not receiving by noon on August 23rd an answer from the Imperial German government signifying unconditional acceptance of the above advices offered by the Imperial Japanese government, Japan will be compelled to take such action as it may deem necessary to meet the situation."

When no answer had been received at the expiration of the appointed time, August 23, the Japanese government declared war on Germany.

We turn our attention to another party playing a secondary rôle in the great drama, this time an ally of the Teutonic empires.

The entrance of the Ottoman Empire into the struggle as an ally of Germany and Austria-Hungary, although apparently justified by no provocation on the part of the allied powers of the opposing group, may be regarded as the natural and almost inevitable outcome of tendencies which have been described in the fifth chapter. It may eventually come to be considered a fortunate occurrence, by making the scope of the final settlement at the termination of the war more inclusive. The Teutonic powers counted from the first, no doubt, upon the participation of Turkey as their ally in the war, and contemplated with eagerness, very likely, the prospect of the proclamation of the Jihad, or holy war of the Moslems against the foes of Islam, with the terror-inspiring effect which that would presumably have on their opponents.

Turkey's military power had been very much curtailed in the rude buffeting inflicted upon her during the Balkan War. She had lost great quantities of military arms and stores. She was probably not able to put more than 125,000 trained soldiers into the field, and her fleet was unimportant and poorly manned. The Ottoman Empire still embraced 694,860 square miles and contained 20,180,000 people. Its great advantages were its incomparable strategic position, and the boundless but vague possibility of inflaming the fanaticism of the millions of Mohammedans living in the dominions of the *entente* allies, in India, Russia, Egypt, and French northern Africa. We have noticed that the railway system, as planned chiefly by Germans, was admirably adapted for the disposition of troops either for offensive or defensive purposes. Railway communication with the Teutonic empires remained open by way of Bulgaria and Roumania.

The dreams of Pan-Islamism for the aggrandizement of the Sultan's empire were supported by the solid facts of history. During at least four different periods the world's greatest commercial center had lain within the territory which the Sultan still possessed or claimed. In remotest times Babylon and Nineveh were the converging points of the great trade routes. In a later period Alexandria, near the mouth of the Nile, was the teeming metropolis of commerce, industry, and wealth.

Then, after an interval of several centuries, the star of commerce hovered over Constantinople, whose opulence cast its spell over the great barbarian world of the Slavs. And, finally, before the glory of Constantinople had been eclipsed, Bagdad made herself the central point in the vastly remunerative trade of east and west. Turkey in the twentieth century thought that she saw wonderful opportunities almost within her grasp, and the enticing prospect of the Suez Canal and the possibilities of Egypt must have been the most stimulating incentive to the more ambitious spirits in the Turkish nation. Undoubtedly the recovery of Egypt was the most cherished aspiration of the leaders of the Pan-Islamic movement, and with this project their German friends would be heartily in accord. The seizure of the Suez Canal would be a most effective blow at the commercial life of Great Britain and the British Empire. The coöperation of the Ottoman Empire with the Teutonic powers appeared, therefore, most reasonable.

But German calculations were disappointed in some other respects, as we have already observed, and Turkey was only drawn into the war after nearly four months of uncertainty and hesitation. The external occurrences during this period of vacillation are characteristic of the evasive, procrastinating nature of Turkish political dealings with the usual unconsciously humorous exhibitions of plausible gravity as a transparent mask of deception. But behind these outward facts there was a mysterious background of factional intrigue and the conflict of rival ambitions, which would probably prove to be a drama of absorbing interest, if it were ever revealed.

The Turkish ministry whose existence had been ushered in by the *coup d'état* of January, 1913, and the murder of Nazim Pasha, Minister of War, was terminated also by a political murder. Shevket Pasha, the Grand Vizier, was shot and killed by an assassin as he was riding to the Sublime Porte in a motor-car, June 12, 1913. But the Committee of Union and Progress, representing the Young Turkish party, which had recovered its authority by the acts of violence in January, 1913, survived the loss of Adrianople, which it had pledged itself not to give up, survived the

assassination of Shevket Pasha, and received increased prestige by the recovery of Adrianople in the summer of 1913. Said Halim Pasha succeeded Shevket Pasha as Grand Vizier. Germanophile sentiments held the ascendancy in the Committee of Union and Progress, although the Grand Vizier professed to be opposed to Turkey's participation in the war and was regarded by the British ambassador as sincere in his deprecations of German intrigues. But four other members of the cabinet, of marked ability and apparently greater force of character, claim our attention. Djavid Bey, possessing an active, adaptable mind, directed the financial affairs of the empire. Djemal Pasha was minister of the navy. Talaat Pasha, Minister of the Interior, was endowed with tact, poise, intelligence and other valuable qualities of statesmanship. Most conspicuous of all was Enver Pasha, the son-in-law of the Sultan, Mohammed V, who was young, spirited, and ambitious. Enver Pasha's career had been a suitable preparation for playing a popular, heroic rôle in the present crisis. He had been in training as military attaché in Berlin. He had commanded the Young Turkish army which suppressed the counter revolution in Constantinople, he had organized the resistance of the Arab tribes in Tripoli against the Italians, and he had derived the chief credit for the retaking of Adrianople in 1913. Thus his name was associated with thrilling, patriotic actions. At the beginning of the great European struggle he was Turkish minister of war at the age of thirty-five. He undoubtedly aspires to play a great rôle; and we may assume that the assassin of Nazim Pasha is a character of inflexible resolution and unscrupulous as to his means. He is destined to be a factor for good or evil to his nation. His pro-German sympathy was very marked.

The German military commission, with General Liman von Sanders at its head, did not depart from Turkey when hostilities commenced between the Great Powers. The Grand Vizier explained that "the retention of the German military mission meant nothing and had no political significance. As they had offered to remain, it would have been ungracious to refuse." But it appears that German officers were continually arriving, and a German officer was entrusted with the command of the Dardanelles, where the forts received German garrisons. The German ambassador, General Liman von Sanders, and Enver Pasha really had the key to the situation in their hands from the beginning, through the control of the army by the pro-German element, although they probably experienced much difficulty in winning over several personages, perhaps the Sultan himself, to their policy.

From the first there was an obvious tendency to arouse the people by imputing hostile motives to the *entente* powers. On August 3, the British government announced its intention of requisitioning two battle-ships which were being constructed for the Ottoman government at a yard on the Tyne in England. The news of this measure was adroitly manipulated so as to inflame popular indignation. The practice, of which this step of the British government is an example, is not unusual; and full requital was promised for any loss that should be incurred by the Turkish government. Finally, the British government sent a formal communication to the Sultan stating "that the exigencies of the defense of the king's dominions are the only cause of the detention of these ships, which His Majesty hopes will not be for long, it being the intention of His Majesty's government to restore them to the Ottoman government at the end of the war, in the event of the maintenance of a strict neutrality by Turkey." The conviction spread, however, that Great Britain contemplated some unfriendly act, either the annexation of Egypt, or an attack upon Turkey herself.

On August 18, the British ambassador assured the Turkish government that the allied powers would uphold Turkey's independence and integrity against any enemies that might take advantage of the general situation to attack her. But two days later, apparently in reply to this communication, the Turkish government submitted a list of terms to the British ambassador to serve as basis for Turkish neutrality. This proposal included the abolition of the Capitulations, the immediate consignment to Turkey of the requisitioned battle-ships, the renunciation of all interference with the internal affairs of Turkey, the promise of the restoration of western Thrace to Turkey, in case Bulgaria should intervene against the *Triple Entente*, and the restoration to Turkey of the Greek islands whose fate still remained undecided.

Sir Edward Grey did not entirely reject these proposals. He was willing to make them the subject for discussion. On August 22, the *entente* powers offered to give a joint guarantee in writing to respect the independence and integrity of Turkey if she maintained neutrality, and Sir Edward Grey repeated this same offer on September 16.

During all this time the Turkish government was repeatedly protesting its intention of remaining neutral. But evidence was continually accumulating to contradict these protestations and test the forbearance of the *entente* powers. In fact, the Turkish government acquiesced in various actions in violation of most of the rules of conduct incumbent upon neutral powers. The most conspicuous violation of neutrality occurred in consequence of the arrival in the Straits of two German war-vessels.

The battle-cruiser *Goeben* and armored-cruiser *Breslau* were in the Mediterranean at the outbreak of hostilities. After causing annoyance along the coast of Algeria, they were pursued by a superior French and English squadron and took refuge in the Straits of Messina; then, eluding their pursuers, they safely arrived at the Dardanelles, August 10.

The affair of the *Goeben* and *Breslau* assumed the character of a comedy of transparent dissimulation. By international law the Turkish government should have insisted that these belligerent ships depart within twenty-four hours, or else intern, that is, disarm and go out of commission for the duration of the war. But the Turkish government displayed no inclination to enforce such a requirement. The two war-vessels, armed and fully manned with their German crews, continued to enjoy the hospitable protection of the Straits. First, the Turkish government announced that it had purchased the *Goeben* and *Breslau*, adroitly giving out that this measure was necessary, to replace the two battle-ships which had been building for Turkey in an English yard. At this time the Grand Vizier requested that the British naval commission under Admiral Limpus should be permitted to remain, and Sir Edward Grey replied that this would be allowed, if the German crews should be sent back to Germany immediately, so that the two German ships should receive Turkish crews as proof that the transfer had been a genuine transaction. The minister of marine assured Admiral Limpus that Turkish crews would be provided for the two vessels as soon as possible. On the very next day, however, Admiral Limpus and all the other British officers composing the commission for the reorganization of the Ottoman navy were removed from their active positions in the fleet and requested to confine their activities within the ministry of marine. In spite of this palpable slight the British naval commission was not withdrawn until September 8th.

Meantime the German crews remained on the two vessels, and on August 18th the Grand Vizier excused this breach of neutrality by the deficiency in available Turkish sailors. He promised to send off the German crews gradually as soon as a transport came from London with Turkish crews.

Although the Turkish transport arrived on the 22nd the German crews remained and were actually reinforced from Germany. The British ambassador reported evidence, moreover, that the two ships had never been sold to Turkey. On September 21st the Sultan explained that the German crews "had been kept for a short time to train the Turkish crews. The German crews would be sent away in five or ten days, and the officers also." But these solemn assurances were never fulfilled.

In addition to the unfriendly conduct of the Turkish government in harboring the two war-vessels, extensive supplies of war material were constantly arriving from Germany, while the number of German officers and men was increasing. The British ambassador reported his observation to the Grand Vizier on September 20 that "Constantinople and the neighborhood formed nothing more nor less than an armed German camp. Many more German officers and men had arrived, and there must now be between 4,000 and 5,000 German soldiers and sailors here. We all, including his Highness, were at the mercy of Liman Pasha (General Liman von Sanders) and the minister of war."

On September 9 the Grand Vizier announced to the powers the intention of the Turkish government of abolishing the Capitulations, this measure to go into effect October 1. These Capitulations may be regarded as the continuation of concessions to the commercial communities of western Europe made by the rulers of Constantinople before the Turkish conquest. In their existing form they were conventions based upon treaties between the Ottoman government and many of the leading nations, guaranteeing to the citizens of the latter, while residing in Turkey, such privileges as exemption from taxation and jurisdiction in their own consular courts, and imposing a limitation upon the percentage of customs duties levied upon imported merchandise. The Capitulations were a restriction upon the full exercise of sovereignty by the Turkish government. Their existence implied the assumption on the part of the western powers that the persons and property of their own citizens could not safely be entrusted to the unlimited authority of the Turkish government.

They were the visible mark of political inferiority, which was galling to the more spirited element among the Turks, those who aspired to see their country occupy a respected position among the progressive states. Ever since the institution of the constitutional régime the Young Turks had cherished the vision of securing the abrogation of the Capitulations as the crowning step in the recognition by the greater powers of the political equality of the Ottoman Empire. In addition to the detriment to their national pride, the financial limitations imposed by the Capitulations may very likely have occasioned a practical embarrassment to the Turkish government. For the immunity of foreigners had as natural counterpart a greater burden of taxation for Turkish subjects, and in spite of the universal tendency for the financial requirements of governments to increase, the Turkish government could not augment the rate of import dues without the consent of the great commercial powers. However, the nations protested generally at this high-handed, one-sided method of annulling treaties. The choice of an occasion for this action by the Turkish government, a time when most of the Great Powers were too much absorbed in a struggle for existence to be able to divert much attention to relatively secondary matters, excited a feeling of annoyance among most of the nations whose privileges were thus abolished.

Although the German ambassador disclaimed authorship for this move of Turkey, and joined in the protest of the powers against it, the belief is quite commonly entertained that Germany really encouraged Turkey to take this defiant step, hoping to involve her in this way in a conflict with the *entente* powers. But the latter exhibited the same forbearance as in the *Goeben* and *Breslau* affair. Sir Edward Grey announced that if Turkey maintained her neutrality, the British government would "be prepared to consider reasonable concessions about Capitulations." The powers of the *entente* may have believed that a conciliatory attitude regarding this question might strengthen the position of the peace party in Turkey, of which the Grand Vizier was represented as being the leading spirit.

The Turkish government had ordered a partial mobilization as soon as the European war had been announced, which they explained as a precautionary measure against any hostile movement on the part of Bulgaria. But as early as the end of August there was evidence of a contemplated attack on Egypt. Extensive movements of Turkish troops in the direction of the province of Hedjaz were reported. On September 23, in making representations to the Grand Vizier concerning the extraordinary concentration of Turkish troops in Syria, which seemed to threaten the security of Egypt, the British ambassador observed that the proceedings in Turkey might afford Great Britain ample justification for protesting against the violation of neutrality. The fact that she had not taken such action was to be ascribed to her hope that the peace party would win the day.

During this time numbers of German officers were being sent to Syria to superintend the preparation of troops and the concentration of supplies. Evidence collected in various quarters seemed to reveal a very widespread Turco-German campaign of anti-British propaganda, emissaries having been sent to India, Yemen, Senoussi, and Persia, as well as to Egypt for this purpose. The belief was systematically disseminated that Great Britain was the enemy of Islam; and the Turkish press, although subjected to rigid censorship, was a very active auxiliary in this undertaking. Thus the wildest reports were published tending to establish the hostility of Great Britain to the Mohammedan religion, or her military weakness, or the rebellion of Mohammedan nationalities against her domination.

As late as October 23 the Grand Vizier and other ministers were still protesting their pacific intentions. On that day the British ambassador learnt that large quantities of gold had arrived from Germany and Austria; and, as the beginning of active hostilities followed only a few days later, it was a quite natural inference that Turkey's intervention was dependent upon financial assistance from the Teutonic empires.

On October 24 Sir Edward Grey instructed the British ambassador in Constantinople as follows:

"You should enumerate to Grand Vizier the hostile acts of which we complain, and warn him that, if German influences succeed in pushing Turkey to cross the frontiers of Egypt and threaten the international Suez Canal, which we are bound to preserve, it will not be we, but Turkey, that will have aggressively disturbed the *status quo*. The following is a convenient summary of Turkish acts of which we complain and which, combined, produce a most unfavorable impression. You might send it to Grand Vizier:

"The Mosul and Damascus army corps have, since their mobilization, been constantly sending troops south preparatory to an invasion of Egypt and the Suez Canal from Akaba and Gaza. A large body of Bedouin Arabs has been called out and armed to assist in this venture. Transport has been collected and roads have been prepared up to the frontier of Egypt. Mines have been dispatched to be laid in the Gulf of Akaba to protect the force from naval attack,

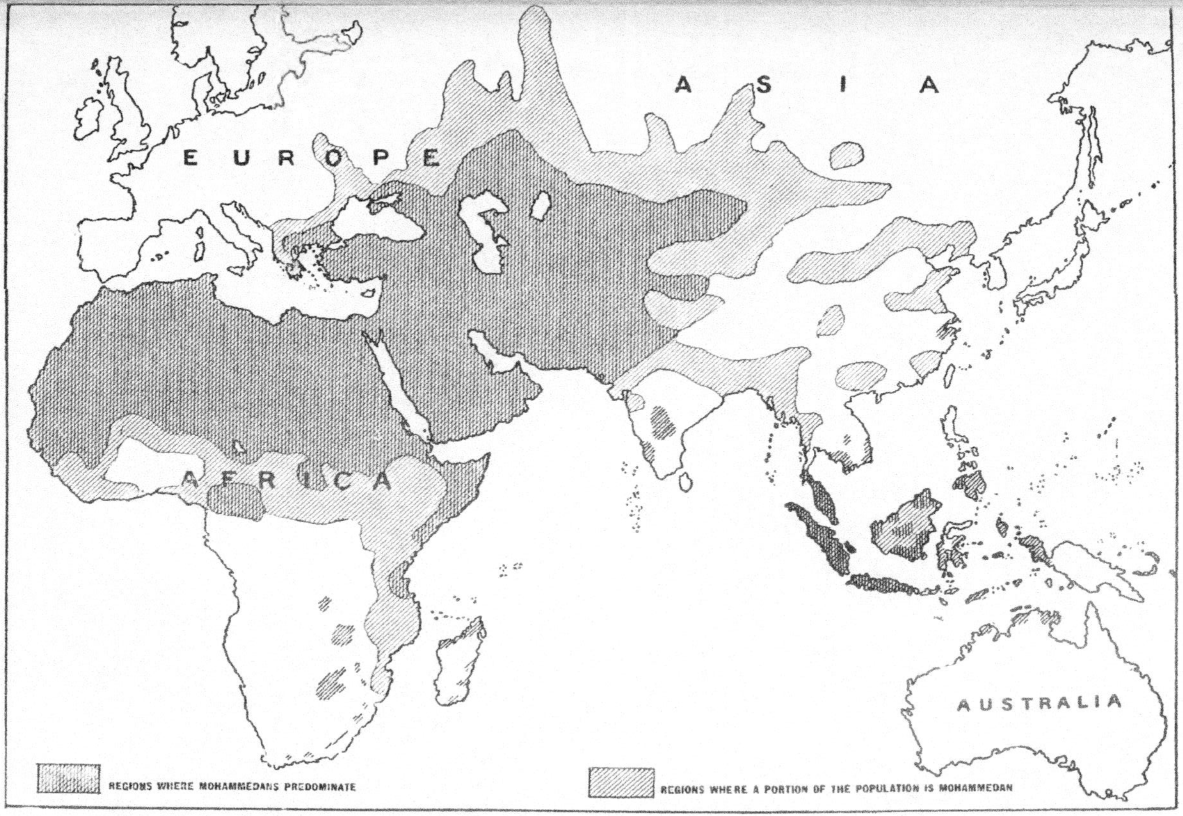

Map showing the proportion of the earth's surface inhabited by Mohammedans, who were expected, on the declaration of a Holy War by Sheich-ul-Islam, to take up arms.

and the notorious Sheich Aziz Shawish, who has been so well known as a firebrand in raising Moslem feeling against Christians, has published and disseminated through Syria, and probably India, an inflammatory document urging Mohammedans to fight against Great Britain. Dr. Prüffer, who was so long engaged in intrigues in Cairo against the British occupation, and is now attached to the German Embassy in Constantinople, has been busily occupied in Syria in trying to incite the people to take part in this conflict."

But on October 26, a force of 2000 armed Bedouins advanced to a point twenty miles within the Egyptian frontier on their way to attack the Suez Canal; and actions taking place on October 29 definitely closed the long period of uncertainty as to Turkey's eventual conduct. Before dawn Turkish torpedo-boats raided Odessa harbor, and bombarded the town, and on the same day the *Breslau* and *Hamidieh* shelled other Russian places along the Black Sea coast. According to Turkish reports, on the other hand, Russian war-vessels opened fire without provocation on a Turkish fleet maneuvering in the Black Sea, and in the ensuing naval combat the Russians lost the mine-layer *Pruth*, which had been sent to strew mines at the entrance of the Bosphorus.

On the afternoon of the 30th the Russian, French, and British ambassadors at Constantinople asked for their passports. Official notice of the existence of a state of war between Great Britain and the Ottoman Empire appeared in the *Times*, November 5, and at the same time the British government proclaimed the annexation of Cyprus. Great Britain had "occupied" and administered this island since the time of the Treaty of Berlin in a manner analogous to the position of Austria-Hungary in Bosnia and Herzegovina.

In a speech at the Lord Mayor's banquet at the Guildhall,

November 9, the British Prime Minister, Mr. Asquith, undertook to explain that the Allies were not making war on Islam, but the political dominion of the Turkish government. The most significant passages in his speech are the following:

"It is not the Turkish people, it is the Ottoman government that has drawn the sword, and which, I do not hesitate to predict, will perish by the sword. It is they and not we who have rung the death knell of the Ottoman dominion, not only in Europe, but in Asia. With their disappearance, at least, will disappear, as I at least hope and believe, the blight which for generations past has withered some of the fairest regions of the earth.

"We have no quarrel with the Mussulman subjects of the Sultan. Our sovereign claims amongst the most loyal of his subjects millions of men who hold the Mussulman faith. Nothing is further from our thoughts or intentions than to initiate or encourage a crusade against their creed. Their holy places we are prepared, if any such need should arise, to defend against all invaders and keep them inviolate.

"The Turkish Empire has committed suicide, and dug with its own hands its grave."

Consistently with the erroneous notion that the Mohammedan religion was being attacked, which had been so diligently cultivated, the Jihad was proclaimed by the Sheich-ul-Islam, the spiritual head of the Mohammedan religion. This is a holy war for the extension of the religion of Islam, which usually offers the victims the choice of conversion or the sword or submission to a heavy tax. Under some circumstances the non-believer is offered no alternative for extermination; and in the present Fetwa, or decree, there is only reference to the killing of English, French, and Russians and the seizure of their property. This decree expressly declares it to be the duty of the Mohammedan subjects of

Great Britain, France, and Russia to wage the Jihad, and anathematizes any Mussulman who might serve in the armies of the Allies against the powers friendly to the Turkish Empire.

It may not be inappropriate to observe, by way of anticipation, that the effect of the Jihad was disappointing to those who had planned it; for it was received with comparative indifference throughout the Ottoman Empire, and was probably even more barren of results in the Moslem lands elsewhere. This was partly due to the decadence of the old spirit of religious fanaticism. Then, the causes of the war were purely political, and this could not be entirely concealed from the Mohammedan population. It was fundamentally a struggle between different groups of Christian states, in which the Mohammedan religion was neither in any way directly involved nor threatened. Perhaps also, the jealousy of the Holy City, Mecca, for the upstart religious authority of Constantinople and the impatient spirit of the Arabs under Turkish domination are partially accountable for the diminished effect of the decree.

In analyzing the motives for Italy's much-debated entry into the field of hostilities we must traverse a long stretch forwards along the chronological pathway of events. The hostility of Italy for Austria-Hungary was due to sentimental, strategic, and economic causes. But the process by which this hostility was translated into action was a deliberate one.

Allusion has already been made to *Italia Irredenta*, Unredeemed Italy. This expression with its special, local significance has become so familiar that a term of universal application has grown out of it; and so, irredentism in the general sense may be defined as a movement to secure the union with a national state of contiguous territories inhabited by people of the same race and speaking the same language.

About 800,000 Italians were living in the regions of the Austrian Empire adjacent to Italy at the outbreak of the European war, chiefly in the southernmost part of Tyrol about Trent, the district known as the Trentino, in Trieste, and in Istria. There were 373,000 in the Trentino; three-fourths of the 161,000 inhabitants of Trieste were Italians, the rest being mainly Slavs. The coast of Istria was Italian, while in the interior Slavs predominated. It is hardly necessary to add that these Austrian subjects of Italian nationality desired political union with the Kingdom of Italy, and that in many ways the attitude of the Austrian government by its lack of tact had intensified their yearning.

Italy's unfavorable strategic position was a cause of continual preoccupation. The general character of her extensive coast-line presented an unusually difficult problem of defense, which was greatly aggravated by the special features of the situation to the east. The balance of power in the Adriatic was vital for Italy's safety, and this balance was chiefly determined by the state of affairs on the eastern shore of the sea. For the Adriatic is controlled in a naval sense by the important positions on its eastern margin. The Adriatic coast of the Italian peninsula is almost without suitable harbors to serve as naval bases; but along the eastern shore of the same sea the condition is quite the reverse. Pola in the north and Valona in the south are ports of inestimable naval importance; and between these two the islands of the Dalmatian archipelago form a veritable labyrinth of sheltered, navigable passages. The greater part of the coast-line is therefore a continuous harbor. In ancient times the Roman conquest of the Illyrian coast, as this shore was then called, the nest of piratical tribes, the forefathers of the present Albanians, was a necessary measure for the security of Italy. Later the possession of Dalmatia became a necessary guarantee for the commercial supremacy of

Venice, and the vestiges of the proud republic of St. Mark are still in evidence along the entire coast. At present the possession of Pola and Valona by the same great power would destroy the independence of Italy; and therefore, since Italy has not been, heretofore, in a position to insist upon the annexation of Valona, she associated herself with Austria-Hungary in the plan of establishing an independent Albania, so as to eliminate that part of the coast from the possible elements forming the balance of power between her ally and herself.

On Italy's northeastern frontier the course of the Austro-Italian boundary was favorable to Austrian military interests, since it left the dominating positions in Austrian hands, and marked out Southern Tyrol in the shape of a wedge driven far into Italian territory, a sally-port from which the Dual Monarchy might at any time launch a mortal blow.

The strategic cause for Italy's hostility was, therefore, her failure to possess natural, strategic boundaries and the fact that Austria-Hungary everywhere blocked the road to their attainment.

During the last fifteen years the development of Italian industry has been increasing in velocity at a very satisfactory, if not astonishing, rate. The possession of extensive sources of power in the waterfalls, rendered available by the progress of electrical invention, the presence of a numerous, industrious, laboring class, and the admirable situation of the country for purposes of commerce have all contributed to this happy result. With scarcely any native deposits of coal or iron Italy supplies her own steel rails and electrical machinery, and competes in the markets of the world in the products of her textile industries, which gave employment in 1911 to 657,190 persons. The Balkan peninsula and Asia Minor constitute a very important market for Italian products; but precisely in these regions Italy encounters the stiffest competition of Austria-Hungary, and the latter's rivalry is rendered especially formidable by the policy of her government to establish a kind of commercial hegemony by means of political and diplomatic pressure.

During the forty-four years of her national existence since securing Rome as her capital, the oscillation in Italy's foreign policy has been determined by the balance of the conflicting sentiments of apprehension and hostility for France and Austria-Hungary respectively.

In the early years of the Third Republic, when the conservative or reactionary elements were most influential in France, the temperament of the French government kept alive the suspicion in Italy that Napoleon III's policy of upholding the papal claim to secular authority would be reasserted. The feeling of apprehension respecting the designs of France rose to a climax with the French annexation of Tunis, when Italy united with Germany and with Austria-Hungary, her former enemy, in the Triple Alliance, as a measure for self-protection. After this the Italian government was naturally compelled to adopt an attitude of formal disapproval of Irredentist yearnings; and, in fact, the spirit of Irredentism seems to have waned for more than twenty-five years.

But in recent years the balance of feeling in Italy regarding the neighboring powers has undergone a remarkable alteration in consequence of the evolution of political tendencies in France and Austria-Hungary. Thus in France a spirit of violent hostility to the influence of the church had supplanted the former ultramontane sympathy, whilst clericalism had become so strong in the court of Vienna that Francis Joseph never paid a visit to the royal court of Rome for fear of offending the Vatican.

Germany and Austria-Hungary had adopted a more energetic attitude in their foreign relations, for which they could not claim the support of Italian interests, as was apparent at the Algeciras Convention. But Italy's attitude of indifference was converted into one of suspicion and jealousy with respect to the policy of Austria-Hungary when a succession of events seemed to afford proof of a design of extending the Hapsburg sphere of influence in the Balkan peninsula.

When war between the central powers and Russia and France became practically inevitable, Italy assumed precisely the attitude which was to be expected. The Marquis di San Giuliano, then Italian Foreign Secretary, informed Herr von Flotow, the German Ambassador, that as the war undertaken by Austria, especially in view of its possible consequences, had an aggressive character, and was not in accordance with the purely defensive purpose of the Triple Alliance, Italy could not take part in it.

Professor Hans Delbrück affirms that if Great Britain had remained neutral, Italy would have taken her stand by the side of her associates in the Triple Alliance; but in view of the general trend of events this scarcely seems probable. It may be remarked, moreover, that the statement of the Italian foreign secretary was made on July 31, whereas Great Britain took no step indicating an intention of intervening against Germany before August 2.

No considerable party in Italy advocated military coöperation with the Teutonic empires. The field for discussion was practically reserved for the controversial combat of the Neutralists and Interventionists, the latter supporting an active combination with the *entente* powers. An overwhelming majority of the nation were in favor of peace, but at the same time an overwhelming majority were determined that some concession must be made toward the satisfaction of Italy's national aspirations.

A very brief survey of some of the salient features of

Italy's internal political situation will aid us in comprehending the influences which were effective in molding the foreign policy of the kingdom in this critical period.

Formerly the Conservatives and Liberals, or the Right and Left, were distinct, and individually fairly coherent, political parties. But these parties have gradually disintegrated into smaller groups. A very large majority of the deputies in the Chamber at present call themselves Liberals; but the Liberal party as such no longer plays any distinctive rôle. The various groups into which the older parties have been dissolved are frequently made up of the adherents of prominent political leaders. But some observers in Italy have attributed this evolution to the rising political conscience of the nation and the consequent revolt from the thraldom of party. Undoubtedly the progress of socialistic doctrine has been an important factor in the transition; and it may be remarked, as proof of the general progress in ideas, that views which were regarded as subversive socialistic heresies in 1898 had become a part of orthodox liberal doctrine in 1914.

Signor Giolitti was the dominating spirit in Italian politics for more than ten years. In political integrity and sagacity he occupied a relatively respectable position among the offspring of an unheroic age. He combined discretion in policy with the popularity and adroitness which secured for him a position of almost dictatorial authority. Under his influence the electoral law of 1913 was passed, which increased the number of voters from 3,500,000 to 8,500,000.

Giolitti withdrew from the premiership in the spring of 1914 in consequence of a defection of the Radicals. He was succeeded by Signor Antonio Salandra, who was at first believed to be nothing more than Giolitti's creature or lieutenant, but who in reality, as we shall presently observe, has displayed a remarkable degree of independence.

Marquis di San Giuliano,
Italian Minister of Foreign Affairs at the outbreak of the war.

Victor Emanuel III,
King of Italy.

Sydney Sonnino,
Italian Minister of Foreign Affairs.

The Italian Chamber of Deputies during the first year of the war was the first one elected on the basis of the new electoral law of 1913. The supreme question of peace or war brought about a curious alignment of parties in this assembly. For the Socialists and Clericals ranged themselves together in favor of neutrality, while the Reformed Socialists, Republicans, and Nationalists advocated active participation in the war on the side of the *entente* powers. The Nationalists are a group of recent formation. The expansion of national ambition encouraged by the conquest of Tripoli nourished the so-called "nationalist" movement, the program of which was at first extensive but vague. The outbreak of the European war naturally focused the attention of the Nationalists on *Italia Irredenta*. As a matter of fact, the Nationalists are recruited from all the parties and are but loosely united.

Signor Giolitti made himself the leading advocate of continued neutrality. He believed that Italy was not ready for war, and that she could obtain adequate concessions from Austria-Hungary without abandoning her attitude of neutrality.

One circumstance of importance for the general situation may as well be mentioned here, without implying any immediate connection with the context. German investments in Italy have increased very rapidly of late years, and are said to amount in value at the present time to about $600,000,000.

The Marquis di San Giuliano, who had been foreign minister since 1905, and had always been favorable to the maintenance of the Triple Alliance—it had been renewed in 1912 during his tenure of the foreign office—died on October 16, 1914. His successor was Baron Sydney Sonnino, who had formerly been leader of the parliamentary opposition to Giolitti.

It is necessary at this point to mention an event which

was not without an influence on the international situation in relation to Italy. After a troubled reign of about eight months William of Wied issued a rather ambiguous proclamation to his Albanian subjects in August, placed the royal seals on the doors of the palace in Durazzo, and departed for Germany, where he shortly joined his old regiment in the war against France, and finally abdicated by formal announcement the throne of Albania.

William's life as ruler of Albania during several months must have resembled that of Otho, King of the Hellenes, whom Thackeray represents as going "out to drive (revolutions permitting) at five." The prediction made by the same authority might just as well have been applied (*mutatis nominibus*) to Durazzo and William: "It was a farce to make this place into a kingly capital; and I make no manner of doubt that King Otho, the very day he can get away unperceived, and can get together the passage-money, will be off for dear old Deutschland, Fatherland, Beerland!"

There were halcyon days in Rome in the spring of 1914 when one could lounge at one's ease in the cafés and experience a varied round of fresh sensations from the daily accounts of stirring events in Albania; treason, conspiracy, insurrection, coups d'état, royal flights, which lost nothing by their narration in the pages of the *Messagero* or the *Corriere della Sera;* all conducted on a scale commensurate with actions in famous Pumpernickel, and involving vastly more actual danger than the court intrigues in that worthy capital. While these premonitory ebullitions in Albania served for the time as a harmless, but not too remote source of entertainment, one little suspected the formidable eruption which impended over Europe.

At the outbreak of the general conflict the powers withdrew their detachments from Scutari. Albania was abandoned to her own turbulent factions, and Prince William found himself without resources or support. There was nothing left for him except to depart as gracefully as he could. In October, Essad Pasha perceiving the opportunity for which he had apparently been waiting since the capitulation of Scutari, hurried to Durazzo at the head of his followers and had himself chosen head of a provisional government by the Albanian senate.

Albanian affairs remained unsettled, and the Italian government sent out an expedition which occupied Valona. It is safe to predict that, as in the case of the Dodecanese, much time will elapse before conditions permit the withdrawal of the Italian troops from Valona.

With the aspirations of Italian Irredentism in mind, it is natural to assume that the Austro-Italian crisis created by the war was a very simple one; that Italy regarded the European conflict as a favorable opportunity for satisfying her national ambition by completing the process of unification, and demanded the cession of the Italian districts in the Austrian Empire as the price of her neutrality. It would follow from this assumption that the conduct of Italy could only be justified if judged on the basis of a standard of political ethics transcending the literal provisions of treaties and the formal rules of international law; in other words, that the case for Italy would rest solely upon the ideal Principle of Nationalities, and the revolutionary doctrine that the exercise of sovereignty in defiance of popular inclination, as in the Trentino and Trieste, is tyrannous and *ipso facto* indefensible.

But the actual situation revealed by the official correspondence is different. For the fact is, that throughout their entire course, the negotiations between the two governments were conducted on the basis of a particular article in the secret treaty which regulated, in a very important respect, the relationship of Italy and Austria-Hungary within the

Triple Alliance. Italy took her stand upon the formal terms of an international agreement.

Article 7 of the treaty referred to provided for the maintenance of an equilibrium between the two allies on the Adriatic and Aegean Seas and in the Balkan peninsula by establishing that, if either should acquire territorial or other advantages in the regions mentioned, such acquisition, even though temporary, would give the ally the right to compensation. It provided, moreover, that the nature of the compensation should be settled by previous accord between the two parties. Thus if Austria-Hungary, for instance, proposed to undertake operations in the Balkans from which any advantage whatsoever might be derived, the treaty placed upon her the obligation of inviting Italy to an exchange of views for the purpose of settling the question of immediate compensation. And so, like a hunter, Austria-Hungary was obliged to settle for the license before taking her chance at the game. But this limitation, which was reciprocal, must not be regarded as an instrument for extortion. It was an indispensable guarantee for the maintenance of the Triple Alliance. The observations already made have shown how the security of Italy was involved in the balance of power to the eastward of the Adriatic. Even with the protection of this guarantee, the Italian government felt that the situation demanded unceasing vigilance; for there was manifest danger that Austro-Hungarian influence might advance in a manner so subtle as to evade the formal stipulation.

It may appear surprising that the interpretation of an article so short and specific could have afforded matter for negotiations lasting more than four months. We shall discover, however, that this famous article 7 left a considerable margin for controversy. Discussion turned upon the questions of whether the actual situation called for compensation, where the compensation was to be provided, when it was to be delivered, and what was to be its extent. The course of the negotiations may be divided roughly into three general periods; the first extending from December 9 until December 20, 1914, before Austria-Hungary had admitted that the actual situation called for compensation; the second, when the question of the location of the compensation was uppermost, from the intervention of Prince von Bülow, December 20, 1914, until the submission of definite proposals by Austria-Hungary, March 27, 1915; and the third, when the principle of a cession of Austrian territory had been admitted, from the latter date until Italy's repudiation of the Triple Alliance, May 4, 1915.

Baron Sonnino, Italian Foreign Minister, inaugurated the discussion, December 9, 1914, by instructing the Italian ambassador in Vienna to inform Count Berchtold that the invasion of Serbia by Austria-Hungary must necessarily be the subject for examination by the Italian and Austro-Hungarian governments on the basis of article 7 of the treaty. Baron Sonnino represented that Italy has a foremost interest in the preservation of the independence of Serbia, and that the Dual Monarchy had violated article 7 by her invasion of the Slav kingdom without previous agreement with Italy. He called attention to the fact that the said article covered temporary occupations of territory in the Balkan peninsula, and even gave Italy the right to compensation for advantages to Austria-Hungary of a non-territorial character. The situation demanded a satisfactory compact between the two powers, and this would invest their alliance with the mutual feeling of cordiality which was desired by both parties.

When the matter was presented in this light to Count Berchtold, he observed that in the varying course of military operations the Austro-Hungarian forces had evacuated the places which they had previously occupied, and that it was manifestly preposterous to demand compensation for an occupation of such a transitory character. When the Duke of Avarna reminded him of the presence of Austro-Hungarian troops in Belgrade, he represented that they also might soon be compelled to evacuate the position which they were holding. The Italian ambassador declared that article 7 granted Italy, unequivocally, the right to compensation for the mere invasion of Serbia, since that tended to disturb the balance of power in the Balkan peninsula. Austria-Hungary, as proof of the justice of Italy's position, had invoked article 7 at the time of the Italo-Turkish War to restrain her ally from executing certain military and naval operations against European Turkey. Count Berchtold asserted that the two cases were not analogous. The proposed action by Italy during the war against Turkey would have been actually aggressive in its nature, and would have endangered the existence of the Ottoman Empire. The present operations of Austria-Hungary, on the other hand, were really defensive in character. This conversation constitutes an interesting revelation respecting the diplomatic situation during the Italo-Turkish War. For the world was informed at the time that Italy's scrupulous abstention from attacking Turkey in Europe was the consequence of her own sense of moderation and of a feeling of loyal friendship for the Ottoman Empire which was proof against every provocation.

The Duke of Avarna insisted that the action of the Dual Monarchy was destroying the equilibrium in the Balkan peninsula (one might have supposed from Austria's protestations in 1913 that there was no equilibrium left to destroy). It appeared that as far back as July 25, 1914, the Duke of Avarna had notified Count Berchtold that Italy would regard as a violation of the Triple Alliance any occupation, however temporary, of Serbian territory, without the required previous understanding with herself. Count Berchtold took cover behind the argument that the presence of Austro-Hungarian troops in positions on Serbian soil in the course of the rapidly shifting operations of a campaign did not constitute even a temporary occupation in the sense of article 7. In fact, a finer distinction must be made between temporary occupations of Balkan territory as covered by article 7 and momentary occupations which are not included within the scope of this provision. Who would hesitate to affirm that the rude spirit of militarism has no place in diplomacy's atmosphere of clear discernment? What soldier "jealous in honor, sudden and quick in quarrel" would have any perception for such subtleties as these?

Baron Sonnino, replying to Count Berchtold's representations through the medium of the Italian ambassador in Vienna, declined to accept the distinction of temporary and momentary occupations of territory. In the Italo-Turkish War Italy had been compelled to abstain, not only from momentary occupations, but even from bombarding points on the coast of European Turkey. The attitude of Austria-Hungary, as based upon article 7, had prolonged the war by prohibiting decisive measures by Italy. It appeared that Count Aehrenthal had vetoed the bombardment of Salonica and the forts at the mouth of the Dardanelles. Baron Sonnino represented, moreover, that the treaty prescribed that the agreement with Italy should be previous to, and not contemporary with, or subsequent to, the military action which made article 7 effective.

On December 20, 1914, the Duke of Avarna reported that Count Berchtold was willing to enter into an exchange of views regarding compensation. On the same day, furthermore, Prince von Bülow, who had come to Rome as German ambassador, and more particularly as special envoy for effecting a reconciliation, had his first audience with Baron Sonnino. On this day the negotiations entered upon their second stage. Prince von Bülow's intervention introduced a new element into the discussion,—Germany's desire for a fair accommodation between her two allies.

Prince von Bülow had accumulated a broad diplomatic experience at Germany's legations and embassies in St. Petersburg, Paris, Rome, and Bucharest, before he was called to Berlin as foreign secretary in 1896. He was very much at home in Italy. His wife is an accomplished Italian lady and he owns Villa Malta in Rome, where many an admiring tourist has gazed at St. Peter's dome, appearing as in a circular composition with perfect balance, through the keyhole of the prince's garden-gate.

Prince von Bülow approached the present situation with his customary affability of manner. He was soon convinced that it would be practically impossible for the Italian government to maintain an attitude of neutrality without obtaining satisfaction for some of the national aspirations.

On January 7, 1915, Baron Sonnino suggested to Baron Macchio, the Austrian Ambassador in Rome, the delicate question of a possible cession to Italy of territories actually belonging to the Austro-Hungarian Monarchy. Baron Macchio made the counter-suggestion that compensation for Italy might be provided in Albania. Baron Sonnino replied that Italy's sole interest in Albania was the negative one that no other power should gain a foothold there. Baron Macchio declared that any agreement regarding compensation would have to be made dependent on the ultimate results of the war. But Baron Sonnino insisted that there must be a fixed minimum of definite advantages for Italy, and he expressed the reassuring conjecture that a self-inflicted amputation of her Italian districts by Austria might eventually contribute to the health and welfare of the

The Bosphorus at its narrowest point, where it is only eight hundred and ten yards wide. *Looking up the straits towards Europe and the Black Sea from a point about seven miles above Scutari. The large buildings on the left, on the European side, belong to the American (Roberts) College; the stone tower with the wall running to the water's edge is the Rumeli Hissar built by the Turks in 1452.*

Dual Monarchy (a worthy object of friendly concern).

On the 14th, von Bülow came forward with his own proposition, which contemplated the cession of the Trentino to Italy, the transaction to remain secret until the end of the war. But Baron Sonnino informed him that the Trentino alone was not enough; Trieste must be added. The German ambassador entreated him not to make the additional demand, because he was convinced that Austria would fight rather than part with her prominent seaport.

A few days later Baron Burian, who had replaced Count Berchtold in the Austro-Hungarian foreign office, suggested that Italy ought to be willing to accept eventual compensation in territory then possessed by Austria-Hungary's adversaries. But Baron Sonnino was of the opinion that such an agreement would be in effect a departure from Italy's neutrality.

Prince von Bülow urged that Italy should herself formulate her requirements; but Baron Sonnino objected that until Austria-Hungary admitted that the discussion was to be conducted on the basis of a cession of Austrian territory, it would be futile for Italy to formulate her demands.

The Duke of Avarna explained to Baron Burian on the 28th that the Italian government was compelled to respond to popular aspirations in directing its national policy. He reiterated the principle that Austria-Hungary's accord with Italy ought to be preventive, and not contemporary with, or consecutive to, the action in the Balkans contemplated by the treaty. He inquired whether the Austro-Hungarian foreign minister accepted the principle that the question of compensation be discussed on the basis of a cession of Austrian territory. Baron Burian replied that he accepted the general principle of a compensation for Italy, but that the cession of Austrian territory was a very serious question requiring further reflection by the competent authorities of

Austria and Hungary separately as well as conjointly.

On February 9, Baron Burian advanced the counterclaim that the treaty imposed upon Italy the obligation of a preventive agreement for her occupation of Valona and some of the Aegean islands, Rhodes and the so-called Dodecanese. But the Duke of Avarna maintained that Italy had occupied Valona merely as agent for the Great Powers, being the only one whose hands were free to uphold the arrangements of the London Conference, and that the islands were being held pending the fulfilment of the terms of the Treaty of Lausanne by Turkey. Baron Burian abandoned Berchtold's distinction of temporary and momentary occupations, but insisted upon the relevancy of his claim respecting Valona and the islands.

In view of this new departure in the negotiations, Baron Sonnino cited a statement by Count Berchtold, made on May 23, 1912, renouncing the right of compensation for the Italian occupation of Rhodes and the Dodecanese, provided the Italians did not proceed any further. In consequence, Italy had abstained from occupying Chios and Mitylene, renouncing precisely the additional strategic advantage which might have forced Turkey to come to terms at once. By prolonging her occupation of the islands, Italy had rendered an important service to Turkey, since, like Chios and Mitylene, these islands would have inevitably fallen a prey to the Greeks, if not protected by Italian garrisons.

As the Austro-Hungarian government seemed to be prolonging the discussion intentionally, merely to avoid an issue, the Italian foreign office withdrew all its earlier proposals, and Italy took her stand firmly on article 7 of the treaty by declaring solemnly that she would consider any action taken henceforth by Austria-Hungary in the Balkan peninsula without a previous agreement with herself regarding compensation as a violation of the treaty. Italy affirmed that the attitude of Count Aehrenthal in prohibiting her action against the remaining islands of the Aegean or the shore of European Turkey was precedent and full justification for her own position in the present circumstances.

Baron Burian undertook to uphold his claim for an understanding on the subject of Italy's occupation of the islands by asserting that if Count Aehrenthal ever waived the right to compensation it was really in the sense of a postponement of the privilege. He declared, however, that he was going to Buda-Pesth the next day to discuss with Count Tisza the question of an agreement with Italy, and that Austria-Hungary would keep in mind the requirements of article 7 in the event of a renewal of the war against Serbia.

On February 27, Baron Burian advanced the view that the actual existence of a state of hostilities between Austria-Hungary and Serbia must be accepted as influencing the degree of priority of the discussion about compensation. It ought to be sufficient, in these circumstances, that the negotiations should have been inaugurated before the commencement of fresh operations against Serbia. If military action had to be put off until the negotiations were completed, Austria-Hungary would in the meantime be rendered defenseless against the Serbs. The Austro-Hungarian foreign minister insisted that the compensation should be proportionate to the advantages obtained by the Dual Monarchy, and still refused to admit, as a basic principle, the cession of Austrian territory. Later, on March 3, he declared that the question of compensation was not pressing, because Austria-Hungary did not contemplate an immediate resumption of activity in Serbia. Since it was recognized that negotiations must be initiated before military operations were resumed, a limit to delay was set which must be reached automatically, so long as Austria-Hungary continued to be at war with Serbia. The Duke of

Avarna reiterated the Italian view that the conclusion of the negotiations must be reached before operations were undertaken. Baron Burian promised to do all he could to hasten their conclusion, but would not bind himself to put off military operations in the meantime. The Duke of Avarna advised Baron Sonnino that he considered that the attempt to bring Baron Burian to the Italian point of view was hopeless.

Accordingly, the next day, Baron Sonnino formulated the Italian position in the following declarations: that an agreement in accordance with article 7 must be concluded before Austria-Hungary should undertake further military action in Serbia, and that a violation of this requirement would be regarded as an open violation of the treaty; that no arrangement for compensation would be acceptable unless it were based on the cession of territory possessed by Austria-Hungary; that a minimum compensation was requisite for the fact alone of operations against Serbia, without excluding further compensation in proportion to the advantages derived by Austria-Hungary; and that the compensation must be immediately effective.

On March 27 we arrive at the commencement of the third and final stage in the negotiations, when Austria-Hungary made her first definite proposal involving a cession of Austrian territory to Italy. On condition of the maintenance of benevolent neutrality throughout the war, an agreement by Italy to grant Austria-Hungary full liberty of action in the Balkans, except in Albania, and the renunciation in advance of any additional compensation for Austro-Hungarian advantages, the Dual Monarchy was prepared to cede territory in Southern Tyrol, including Trent, to Italy. But Austria-Hungary could not admit that it was reasonable to expect that the actual transfer of territory should be effected until the equivalent, Italy's neutrality until the termination of the war, had become a reality.

Baron Sonnino formulated on April 8 the conditions which Italy considered indispensable. They included the cession of the Trentino; the rectification of the Italian boundary in the north-east so that it would pass to the eastward of the Isonzo River, embracing Gradisca and Gorizia within the kingdom, and terminate at the sea near Nabresina; the erection of Trieste and its surroundings into an independent state with its territory extending northward as far as the Italian boundary at Nabresina, and southward so as to embrace a considerable portion of Istria; the cession to Italy of the Curzolari islands off the Dalmatian coast, comprising Lissa, Lesina, Curzola, Lagosta, Cazza, Meleda, and Pelgosa; the immediate transfer of the ceded territory to Italy; recognition by Austria-Hungary of Italian sovereignty in Valona; the relinquishment by Austria-Hungary of her interest in Albania; and the renunciation by Austria-Hungary of the right to demand compensation for the Italian occupation of the Dodecanese. In return for these concessions Italy offered to pay an indemnity of 200,000,000 lire in gold ($38,600,000), to bind herself to neutrality throughout the war, and to renounce any further right to compensation for advantages derived from the war by Austria-Hungary.

Austrian authorities point to these demands as proof that the government at Rome no longer desired a pacific accommodation with the former ally, but had determined to throw in its lot with the *entente* powers. The demands relating to the Trentino, the Isonzo valley, and Trieste might be justified on nationalistic grounds. But the population of the Curzolari islands is almost exclusively Slavic. The attempt to justify the demand for the cession of these islands could be based solely on the strategic argument of securing an outlying bulwark for the protection of the exposed Italian coast.

On the 16th, Baron Burian reported that the demands relative to the correction of the frontier, the city of Trieste, and the Curzolari islands could not be accepted by Austria-Hungary; and on the 21st, Baron Sonnino rejected the Austro-Hungarian proposals. The negotiations came to a deadlock, and all the diplomatic skill of von Bülow was apparently incapable of bridging the chasm which separated the minimum demands of Italy from the maximum concessions of Austria-Hungary.

Accordingly, on May 4, the Salandra ministry took the ominous step of repudiating the Triple Alliance. In announcing its decision the Italian government declared that the alliance, which had been formed as a means and guarantee of peace, had been violated by Austria-Hungary in presenting the ultimatum to Serbia without communicating her intention to Italy or heeding the latter's advice. Austria-Hungary had gone to war to put into execution a program directly opposed to the interests of her ally. The Italian government had striven for months to devise a basis upon which friendly relations might be reëstablished, but all its efforts had been shattered by the resistance of the Austro-Hungarian government, which finally, after many months, had offered an entirely insufficient concession. Italy was accordingly compelled to withdraw all her proposals, resume her entire liberty of action, and declare her treaty of alliance with Austria-Hungary to be void and henceforth of no effect.

Although it was believed that this measure would bring the negotiations to a definite conclusion, another proposal of more comprehensive scope was made by Austria-Hungary, elicited no doubt by an earnest appeal of von Bülow, backed by the approval of the German government. In playing this final card, von Bülow appears to have invoked the coöperation of Signor Giolitti, and to have submitted to his private consideration Austria's final proposal before it was communicated to the Italian government. This proposal embraced the cession of the Trentino and the Italian part of the Isonzo valley, the modification of the political status of Trieste into an imperial free city of the Austrian Empire, the recognition of Italian sovereignty over Valona, and the relinquishment by Austria-Hungary of her interest in Albania and of any claim for compensation arising from the Italian occupation of the Dodecanese. In return Italy was to remain neutral and waive in advance the right to any further compensation for advantages which Austria-Hungary might acquire in the war. The cession of territory was to be effected as soon as practicable. The Italian government made no reply to these proposals.

Signor Giolitti returned to Rome with the intention of casting the weight of his influence into the scales in favor of reconciliation and the maintenance of peace. The question of war or peace had reached its crucial stage. The passionate manifestations in support of the rival policies throughout the country raised popular excitement to fever heat; but it was clear that the balance of popular feeling was now in favor of energetic measures.

The majority of the Chamber, which had been elected when Giolitti was in power, seemed disinclined to support the cabinet's policy of persevering in the demands of April 8. Accordingly, Signor Salandra and his colleagues tendered their resignation to the king.

The supposed interference of Ambassador von Bülow in the field of domestic politics irritated Italian sensibility. Professor Ferrero wrote in the Radical organ, *Il Secolo:*

"Since the time had passed for making new proposals to the ministry, Prince von Bülow entered into an alliance with a group of politicians who were eager to overthrow the ministry and take their place, and of journalists whose services were enlisted in their behalf, and thus succeeded in overthrowing the cabinet for the time being, but not in reversing the current of public opinion. Such methods have been employed by European diplomacy at Constantinople, and at Fez before Morocco was placed under the protection of France. An ambassador who presumed to conduct himself in any European capital as von Bülow has done in Rome should be recalled immediately at the request of the power to which he has been accredited. This formidable crisis ought to be made an object lesson for all the world to show whether Italy will submit to the same sort of treatment as Turkey from German diplomacy; whether she will tolerate the failure to distinguish between Rome and Byzantium."

The passage has been reproduced with a view more especially to illustrate the temper of a large part of the Italian people at this decisive moment as expressed in the words of an individual of international reputation. It is fair to add that no less a personage than the German Chancellor affirms that representatives of the *entente* powers were actively engaged in a campaign of intrigue in Rome at this very time.

Signor Giolitti's intervention did not succeed. His followers dropped away, and the Salandra ministry returned to power in triumph. In a solemn session of the Chamber, May 20, full powers were conferred on the cabinet by a vote of 407 to 74 to carry out their policy, by extreme measures, if necessary.

In the course of an address, which was greeted with tremendous applause, Prime Minister Salandra expressed himself as follows:

"The government has striven patiently through long months to devise some compromise which should restore to the alliance the fundamental basis which it had lost."

When it appeared that the negotiations were being deliberately prolonged to evade the issue, "the Royal (Italian)

FRIEDRICH WILHELM VICTOR ALBERT

William II, King of Prussia and German Emperor.
In field uniform.

government was compelled to announce to the Imperial and Royal (Austro-Hungarian) government, on May 4, the withdrawal of all its proposals, the repudiation of the treaty of alliance, and the declaration of Italy's complete liberty of action.

"It was impossible, moreover, to leave Italy in a position of isolation, deprived of guarantees and prestige, at precisely the moment when the world's history is traversing a decisive stage.........."

"Without arrogance or pride, but with a profound sense of our responsibility in this crisis, we feel that we have satisfied the noblest aspirations and the most vital interests of our fatherland; whose name, with the devotion due it, we invoke in our fervent appeal to parliament, and through parliament to the whole nation, to let all discord cease........

"From to-day onwards, forgetting all other considerations, let us remember only this: that we are all Italians, and love Italy with the same passionate devotion."

It had been regarded as an axiom of diplomacy that Austria and Italy must be either allies or foes, and the events in May, 1915, served to confirm this conviction. For on the 4th, Italy renounced the alliance with Austria-Hungary, and on the 23rd, presented her declaration of war at Vienna.

We pause to determine our impression of the whole series of proceedings between the two former allies. The subject permits no impromptu judgment. It is not an ideally simple conflict, where right and reason are arrayed on one side and falsehood and injustice on the other. Italy, it is true, seems never to have violated the letter of her compact with Austria-Hungary. She may have held her allied rival inexorably to a hard bargain; but Austria-Hungary had become a party to the agreement deliberately, without compulsion, and presumably in full possession of her faculties. She appears to have been satisfied with the bargain for more than thirty years, having renewed it in 1912. We need not, therefore, be more Austrian in our solicitude than the Austrians themselves. Von Bülow's acknowledgment, as early as January, 1915, of Italy's right to a considerable territorial compensation from Austria-Hungary is very strong evidence for the correctness of Italy's attitude in so far as that important point is involved. The weakest element in Italy's position, and at the same time the most difficult feature to judge in the whole discussion, is the question of the extent of Italy's proper compensation, and whether it ought to bear a proportionate relationship to the increase of Austro-Hungarian influence in the Balkan peninsula. After all, we come back to the 800,000 Italian subjects of Austria as a determining factor in this phase of the problem. Should historical title-deeds or popular inclination constitute the more valid claim to sovereignty? This, like some other weighty questions of political ethics, after all the histories have been written, must still be left to the discretion of the individual reader.

In scrutinizing the official correspondence one is apt to be impressed by the seemingly narrow margin which separates the Italian demands of April 8 and the Austro-Hungarian proposal conveyed to the Italian government after its repudiation of the Triple Alliance. The Austro-Hungarian offer failed to cover the Italian demands relating to the Curzolari islands, Trieste and its environs, and a part of the disputed territory in the northeast. But this discrepancy must in reality be reduced by the extent of the value which we ascribe to the Trieste concession and the omission of an indemnity in the Austro-Hungarian proposals. One might be led to suspect that Austria-Hungary's final proposition would have been accepted if it had been offered earlier; and if this is true, Austro-Hungarian diplomacy committed the fatal blunder of misinterpreting the degree of determination in the Italian attitude. After May 4th the question of peace or war in Italy was complicated, and possibly somewhat obscured for a time, in consequence of the intervention of Giolitti, by the political animosity which his name and personality provoked. Besides, the Italian government seems to have concluded a provisional agreement with Great Britain acting for the Allies, on April 25, which defined the terms of Italy's eventual association with the powers of the Triple Entente, in case she declared war against Austria-Hungary before May 25. After May 4 the Italian government very likely felt that their honor as well as safety involved them with the cause of the Allies.

Other considerations support the view that Italy would in no case have accepted the Austro-Hungarian proposals. The formula of April 8 had been offered as an irreducible minimum of Italian pretensions, which the Italian government probably regarded as modest, as falling far below the reasonable expectations of the results of a successful war. The Italians doubtless estimated the concession to Trieste as of small significance; and may have distrusted that, if defeated, the Teutonic powers would be unable to execute all the stipulated points, and if victorious, would discover some pretext for evading them.

CHAPTER X

General Conclusions

Potential causes: Principle of Nationalities not thoroughly realized; problems of Austro-Hungarian populations, and their possible solution; geographic and ethnographic boundaries will not always harmonize; rivalry of Germany and Great Britain; German aspirations and British and French imperialism; war viewed as a biological necessity; the saner German attitude; exaggerated notions relative to foreign commerce, the trade of the United States with China, South America, and Canada furnishing an example; reasonableness of German *Welt-politik;* the development of German sea-power excessive in view of the commercial and political situation; 1911 a turning point. Positive causes: the conflict of the Teutonic powers and Russia in the Balkans; Constantinople the pivotal point; speculation on possible arrangements and the corresponding difficulties; outbreak and spread of the conflagration.

It remains for us to collect and appraise the more important fragmentary conclusions to which the successive stages of our investigation have conducted us. We divided the causes of warfare into potential and positive, and in our examination of the former we discovered that the outward peace in Europe concealed many latent sources of conflict.

Professor Burgess has defined the ideal state as a national unit in a geographic unit, the two coinciding. The nineteenth century carried Europe a long way in the direction of a system of states on such an ideal basis. But the Principle of Nationalities had not attained universal application. Thus the Christian peoples of the Balkan peninsula, in particular, had only partially achieved national independence and unity; the Polish nation remained in a condition of almost complete political effacement; and the very existence of Austria-Hungary seemed to be a defiance of the principle.

The nationalistic aspirations which threatened the integrity of the Hapsburg realm fall into two classes, according as their sources lay outside or inside the monarchy. Thus the movement among the Italians, Serbo-Croats, Roumanians in Transylvania, and Ruthenians tended to detach these populations from the monarchy and unite them with states outside, and the same would be true of Polish ambitions in Galicia, if an independent Poland existed. On the other hand, the extreme nationalistic parties among the Magyars or Czechs aimed to establish greater independence for political units which were included entirely within the Hapsburg realm.

The solution of the Austro-Hungarian internal problems will demand broad statesmanship and an unusual spirit of forbearance on the part of rulers and subjects. The monarchy stands before the supreme trial of its political adaptability. Its complicated situation is a veritable Gordian Knot for political progress, a fateful riddle of failure or achievement, which the rude process of war will probably not succeed in unfolding. We have considered the enlightened policy commonly ascribed to the assassinated Archduke Francis Ferdinand. We have observed that the separatist tendencies within the Hapsburg dominions are really limited in their geographical extent. We conclude that the future security and reputation of the monarchy must be won by a resolute, unprecedented, forward progress, by discarding out-grown convictions, and the encumbering prejudices, jealousies, and deceptive sentiments which tradition has woven about the notions of patriotism and allegiance, and by reërecting the political structure as a federal union, an imperial Switzerland, on the reliable basis of sober reason and popular consent. Unfortunately, traditional habits of thought and emotion, erroneous notions with regard to political values and the proper ethics of sovereignty—in short, a narrow, stupid conception of patriotism, will oppose almost insurmountable barriers to this process. The stubbornness of the Magyars alone, in sustaining their historic rights against the liberation of the Southern Slavs, is a forcible example of the difficulties which would be encountered.

The fact that natural geographic boundaries do not always coincide with the demarcation of nationalities is an unavoidable element of imperfection in any territorial adjustment of European states. Poland, if reconstituted, could have no natural boundaries, except on the south, for, as has been mentioned, the great central Russian plain sweeps right across this territory, and practically includes northern Germany as its extreme projection. The lack of natural barriers between Bulgarians, Serbs, and Greeks in Macedonia was a cause of the second Balkan War. There are instances, moreover, where the position of the ethnic boundary, and consequently its relation to the geographic frontier, is uncertain. The Vosges Mountains, as has been indicated, are the natural boundary between France and Southern Germany; but the French people never became entirely reconciled to the annexation of Alsace by Germany, although it lies on the eastern, or German, side of these mountains. Who, moreover, can accurately define the nationality of the Alsatians? Russia seeks the Carpathian Mountains as her natural geographic boundary towards Austria-Hungary; but it follows by no means that the population intervening between her *ante-bellum* frontier and the crests of this conspicuous range would assimilate with the Russian people. The Polish population would directly associate with their brethren who are already included in the Russian Empire, adding strength to a tenacious national element whose feeling of spiritual community with the Russians has scarcely proceeded beyond a visionary stage.

The most absorbing factor in the international situation in western Europe has been for a number of years the rivalry between Germany and Great Britain. The Germans as a people felt a legitimate pride in their national vitality and capacity. The more intelligent and leading classes were animated with a feeling of assurance in contemplating the past achievements and actual efficiency of the nation in nearly all the lines of human activity which lead to profit and glory. Germany was a veritable storehouse of concentrated energy, and her institutions, by means of their skilful adjustment and effective coöperation, were a marvellously effective system of machinery for regulating and applying this energy. Far from relaxing the national vigor, the rapid accumulation of riches, which was still a quite recent phenomenon, stimulated and mobilized the laborious exertions of the people by exhibiting the prizes of industry and furnishing a more abundant banking capital. German enterprise spread to all parts of the world, and such institutions as the Deutsche Bank became synonymous with the dauntless march of the nation's economic forces.

In the confidence of their youthful vigor, the more impetuous elements in Germany regarded the Russians as decadent before reaching maturity, the English as effete and nerveless, sinking into the slough of selfish commercialism and self-indulgence, and the French as doomed to gradual extinction. They compared their own inadequate territorial allotment with the vast patrimony of their unworthy rivals. The British Empire extended to every quarter of the globe and contained immeasurable opportunities for development. The French colonial empire had been extended under the Third Republic to nearly nine times its extent and eight times its population. But the British Empire had been largely obtained by good luck, and the British nation had never exhibited the disposition to make the far-reaching sacrifices by which military and imperial supremacy is truly merited; whilst the French had so declined in vitality that they could scarcely sustain their population at home, not to mention colonizing extensive new areas.

It cannot be denied that comparisons of this sort were calculated to arouse a feeling of restlessness and dissatisfaction. Scholars and professors proclaimed that the forces of evolution must provide for the survival of the fittest in political organisms, and that strife was nature's purifying medium, by which a suitable readjustment of the existing unhealthy situation might be expected. It would be immoral to thwart nature's beneficent process. Even ancient historians, it may be remarked, reflected the current preconceptions and aspirations by showing that a professional army, like England's, as distinguished from a universal compulsory levy, as in Germany, had been a main cause for the destruction of Greco-Roman society. "It is precisely political idealism that demands wars, while materialism condemns them," exclaimed von Treitschke, and von Bernhardi affirms that war is a biological necessity.

The more moderate, and probably more influential, personalities among the leaders of public opinion in Germany recognized that the greatest part of the world was already unalterably allotted, and that it was futile for them to expect to obtain further extensive colonial territories for themselves, that the brightest hope for their future commercial expansion outside of Europe was associated with the preservation of the "open door" and the economic penetration of such regions as Asia Minor, where Germany could hope to secure a commercial, rather than a political, supremacy.

The political idealism of the Anglo-Saxons—as contrasted at times with their practice—rejects the doctrine of a distinction between public and private morality, that is, that states are not amenable to the same standards of conduct as individuals. It abhors, for instance, the view that disparity in the territorial possessions of nations may rightfully

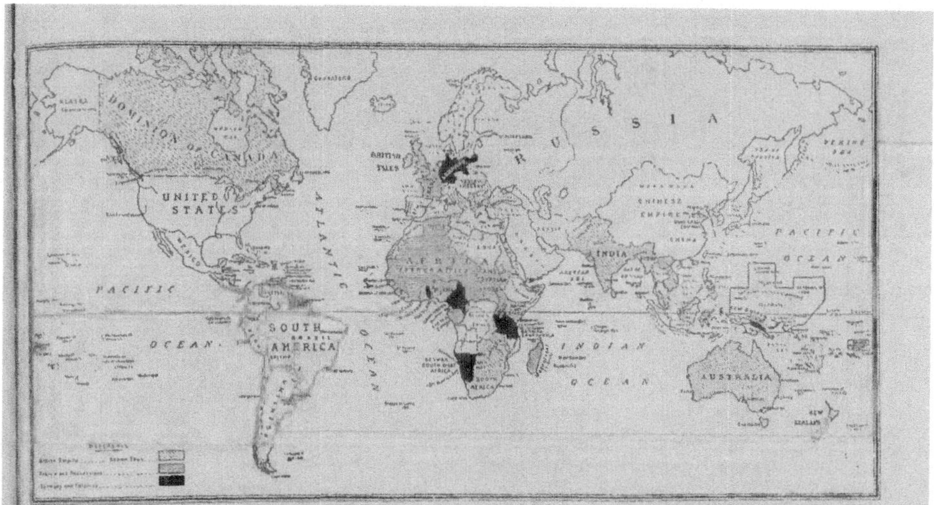

Map showing the relative size of British, French, and German possessions.

be equalized by force. It applies identical notions of justice to the territorial sovereignty of states and the real property of private persons. It regards with composure, for example, the contrast of Japan with her 50,000,000 inhabitants crowded into 140,000 square miles of territory, and Australia with her 5,000,000 people lost within an area of 3,000,000 square miles and refusing to relieve the congestion of the Japanese by admitting them as individuals to the vast spaces which her own people are unable to utilize.

The Anglo-Saxons, as the greatest landed proprietors of the world, are naturally most zealous in their support of established rights; and the assize of the nations accepts generally their attitude, notwithstanding the violations of their own canon, which they themselves in the past have committed.

We observe, however, that even English private law has relaxed somewhat the rigidness of its conception of the sacredness of property, in submitting to such devices as the taxation of unearned increments, the compulsory sale and division of large estates, and other measures intended to restrain the expansion of over-grown fortunes and temper the excessive discrepancies inflicted by chance. And, tempted by analogy, we are led to ask, whether in the world-wide society of nations some similar restraining influences cannot reasonably be applied in a conciliatory spirit to moderate the expansion of the gigantic empires for the benefit of the disinherited, and to mitigate the unfairness of circumstances?

Could not the nations which have been so lavishly endowed be expected to display, if not the spirit of charity, at least the virtue of moderation in mapping out spheres of exclusive political influence beyond the limits of their already unwieldy possessions? And as partial compensation for fortune's shabby treatment of energetic nations like the

German, should not a little more consideration have been displayed, when a fresh division of territory was to be made, and should not the tranquil development of their special commercial interests in the regions like Asiatic Turkey, where no prior rights were being molested, have been cheerfully conceded?

The significance of all observations based on commercial ambitions will be more truly appreciated if we consider them in connection with a characteristic element of exaggeration in the attitude of the age in which we live.

The Kaiser once said that we live in the present age under the star of commerce. If he had wished to define more precisely the charmed influence that captivates the imagination of men, he might appropriately have said the star of foreign commerce, or better, commerce with distant countries. An example will make clear our meaning. The internal commerce of France is estimated to amount in value to ten times her foreign commerce, and this ratio would probably not appear abnormal if compared with the corresponding ratio in many other countries. Yet economists persist in gauging the prosperity of nations according to their volume of foreign trade, regardless of the character of their natural resources and the extent of their dependence on other lands for their necessary supplies. Exports and imports usurp the attention almost exclusively in economic comparisons.

It is the spectacular element in foreign trade which exercises the most potent fascination and even sways the policy of nations to an absurd extent. The memories of the old East India merchantmen, the legendary traditions of the gorgeous riches of the Orient still dominate the fancy; the opulent fertility and alluring opportunities of tropical and remote lands are unquestionably prominent in the counsels of traffic and finance.

It is not necessary for us to go abroad to discover striking examples of this eccentric attitude. For a chorus of voices here in the United States would reëcho the words of Count von Bülow, when he said:

"The Chinese Empire with its population of nearly 400,000,000, is one of the richest markets of the future. We cannot allow ourselves to be excluded from this market, upon which our economic and material progress depends."

Besides this, the possibilities of South America loom ever larger in our imagination. The encouragement of government and press is ready at hand to stimulate a propaganda of exhortation for improving the nation's commercial opportunities in those quarters. An elaborate equipment of steamships, banks, commercial houses, and fleets is demanded for the extension and protection of these interests lying beyond the seas, and "dollar diplomacy" has been designed in their behalf. But in comparison, trade with Canada, which only has to traverse a common boundary, is unheralded and unrenowned. By the side of the splendid allurement of far-away places, the interchange of goods with near-by Canada seems colorless and commonplace. And yet, as regards their profitable character, these three fields for American commercial enterprise suggest the caskets where Bassanio's fortune was concealed. For the "meager lead" of our exportation to Canada amounts in value to fourteen times the "gaudy gold" of our shipments to China, and nearly three times the sum of our exports to all the South American countries; and this contrast, moreover, is increasing rather than diminishing.

This example from right at home is a very palpable illustration of a serious confusion of judgment as to economic values which is common in the commercial nations to-day.

Germany, as we have seen, caught the prevailing infection of inordinate eagerness for this romantic element in commerce; and although her economic development has been characterized by consistency and balance, as could have been inferred from facts already cited, the interest in remoter opportunities has been a very lively factor in shaping the government's policy.

We have discovered that Germany's *Welt-politik*, as described by her leading statesmen, embraced no unreasonable elements. She demanded the preservation of the "open door," wherever it still existed, in other words, that the remaining areas where the nations could compete on an equal footing should not be further restricted, and also that the general balance of colonial possessions should not be disturbed so as to diminish her own relative position.

Ostensibly for upholding these principles and protecting her sea-borne commerce, Germany undertook to create very rapidly a powerful navy; which is a capital fact for the investigation of the potential causes of the war. Nobody denies the right of Germany to construct as large a navy as her resources permitted, but the problem of its expediency inevitably confronts us. The German ministers steadfastly asserted that their program of construction was based solely on a general consideration of the nation's requirements without reference to the naval forces of any other power, although the preamble itself of one of the earlier naval bills suggested a comparison, by affirming that "Germany requires a fleet of such strength that a war against the mightiest naval power would involve risks threatening the supremacy of that power."

The size of the German naval establishment would inevitably be compared with the British in respect to the importance of the interests which each was expected to defend, and the sincerity of Germany's pacific professions would accordingly be judged on the basis of this comparison. The more moderate representatives of German naval policy expressed the opinion that the German navy should maintain the ratio of strength of 10:16 in comparison with the British, and although the government had never made any official recognition of such a standard, it probably answers roughly to the relative naval forces of the two countries in the most recent period. One would not be very far wrong in comparing the extent of the functions to be discharged by the two navies with reference to this basis. Now the colonies and dependencies which required the protection of the German navy were small and unimportant in comparison with those which depended upon the protection of the British navy. The foreign commerce of Germany amounted in value to about three-fourths of that of the United Kingdom. But it must be noted that about one-half of Germany's trade was carried on with countries with which Germany was connected by land routes, and did not directly depend upon the protection of the fleet, while the entire volume of Great Britain's foreign commerce was dependent upon naval protection. Upon the basis of the relative commercial interests requiring protection, therefore, the German navy should have had a strength of only three-eighths ($\frac{1}{2}\times\frac{3}{4}$), instead of five-eighths (10:16) of the British navy. But, most important consideration of all, the British navy was the indispensable guarantee for the very existence of the British nation and the sustenance of the people, whilst Germany could subsist, as the war itself has proved, if deprived of all the interests which her fleet was intended to guard.

It was above all the rapidity with which the German navy had been constructed and was being continually augmented which alarmed public opinion in Great Britain. Germany had been the strongest military power on land for fifty years; and now an additional effort for naval power, an exertion so intense and sustained, seemed capable of but one explanation: that it was intended for a definite, aggressive enterprise. Germany's entire annual expenditure for naval purposes had increased from less than \$40,000,000 in 1900 to more than \$116,000,000 in 1913; Great Britain's, from somewhat less than \$150,000,000 to somewhat more than \$237,000,000 during the same interval. The situation, as presented by these figures, may not appear so threatening for Great Britain, unless we take into account that the Germans were devoting nearly one-half of the annual expenditure to new constructions and had very nearly overtaken the British in this respect. In fact, Germany's annual expenses for naval construction had increased 247% during twelve years, while Great Britain's had only advanced 43%.

The tremendous expansion of the German naval establishment was excessive from both the commercial and political point of view. Germany's best customers were Great Britain, Austria-Hungary, Russia, and the United States. A fleet would be serviceable only for protecting the American commerce. And yet, while German exports to Great Britain had advanced to the astonishing annual value of \$391,320,408 in 1913, having increased about 100% in five years, the German government were persistently engaged in creating uneasiness for their country's best customer by busily increasing their own naval armament, and seemed ever ready to risk this enormously lucrative trade by insisting to the last extremity upon some point involving colonial interests of comparatively slight pecuniary significance, or perhaps only potential advantages.

Whatever was the ultimate intention that actuated the development of German naval power, it produced the effect of a challenge for naval supremacy in driving the British to increased activity and precipitating thereby a wasteful race in naval armaments. Until Germany built her navy, Great Britain held aloof from both the continental groups of powers. To push forward this vast naval program, exciting thereby the antagonism of a great power with which Germany had had no fundamental grounds for conflict, while the continental situation was not free from elements of hostility, may very reasonably impress the observer as an unsound policy. It was the most important factor in breaking down Great Britain's isolation and impelling her to enter into intimate relations with the powers opposed to Germany.

But in the eyes of the public the present war has assumed too exclusively the character of a conflict for commercial supremacy between Great Britain and Germany. A state of tension had existed between the two countries largely in consequence of policies which were intimately related with commercial considerations, it is true, but the immediate, positive cause of the war must be sought elsewhere. It is not likely, moreover, that Germany, France, and Great Britain would have been impelled to hostilities without an incentive from outside. The crest of the wave of warlike peril in the west had been passed at the termination of the Morocco crisis in 1911, and signs were not wanting that relations were becoming much less acute. The year 1911 is, therefore, a turning point. Down to that year the movements and events that portended an approaching universal conflict arrange themselves naturally with reference to the relations of the western powers, but with that year the center of interest is shifted, and henceforth the continuity of events must be traced in the east.

Austria-Hungary had long cherished an interest and aspirations in the Balkans, and Germany virtually indorsed her policy by joining with her in the defensive alliance in 1879. Bismarck, however, shrinking from the dissipation of forces upon erratic undertakings, always insisted that Germany had no direct concern in the regions of the Ottoman Empire. It was part of his discerning policy to maintain an equilibrium in that quarter between the interests of

Austria-Hungary and Russia, whose friendship he cultivated. The entry of Germany in a leading rôle upon the Ottoman stage was, like the creation of the new navy, an achievement of the reign of the present Kaiser. In consequence, the Sultan was gradually reduced to a state of practical subjugation to German influence and commercial ambition. The war between Italy and Turkey and the subsequent critical events drew attention to the prominence of Teutonic interests, although endangering them, and emphasized the essential rivalry of the Teutonic powers and Russia.

Although both these parties professed the conviction that the Balkan states should regulate their affairs independently of outside interference, and perhaps sincerely desired it, the intricacy of their own interests, their extreme sensitiveness to every impression that disturbed the balance, and the resulting condition of nervousness, involved them almost unavoidably in a conflict for supremacy, which was virtually being waged by diplomacy and intrigue during the negotiations and conventions which accompanied the course of the Balkan Wars.

It was desirable in the general interests of progress that Germany should have an unembarrassed opportunity for developing the interior of Asia Minor, and that she should to this end exercise enough political influence with the Porte for assuring the security and welfare of her pioneers of civilization. Yet Russia, like the owner of a great estate, could not remain indifferent as long as her right-of-way from the highroad was subject to the discretion or caprice of a possibly hostile power. The winding water-way through the Bosphorus, Sea of Marmora, and Dardanelles is Russia's tendon of Achilles, the vital artery for her economic life. Constantinople, where the Teutonic line of communication crosses this route, is naturally the pivotal point in the collision of the rival interests.

EUROPE
HEIGHT OF LAND
DEPTH OF SEA
NORTH SEA

An impartial and efficient administration was needed on the Straits, one capable of guaranteeing the security of navigation and traffic. The weak and visionary character of the Turkish government invited the intrigues of the powers, and afforded no assurance of a consistent, independent policy, calculated to inspire confidence. A satisfactory regulation of affairs in the region of the Bosphorus and Dardanelles must be based upon the frank recognition of the supremacy of the major interests of international commerce. An abstract, detached view of the situation might suggest an international condominium, with an administrative commission responsible to the Great Powers as the direct organ of government. The members of such a governing board might suitably be chosen from one or more of the smaller, neutral states, such as Switzerland. But amid the vicissitudes of war, the difficulties of adjusting the views and expectations of allies, and the exigencies of negotiation, so disinterested an arrangement has small chance of realization.

The military problem of defending the Straits is not inconsiderable; for the length of the navigable channel from the entry of the Bosphorus to the point of exit from the Dardanelles into the Aegean Sea is nearly two hundred miles, and the occupation of any one commanding point by an enemy intent upon blockading the passage would render useless the possession of all the remaining course. Besides, the deep indentation of the Gulf of Saros, which exposes the European defenses of the Dardanelles to attack from the rear, is a serious element of weakness. These considerations make it doubtful whether Greece or Bulgaria could defend successfully the extensive position; and it is certain that an arrangement by which one of these Christian states and the Turkish Empire confronted each other emulously on opposite sides of the Straits would offer less security for Russian commerce than that which existed with the ante-bellum situation. All indications point, therefore, to the consummation of the Russian dream of centuries, in the event of Turkish defeat, the control of Constantinople by Russia.

The suspicion and rivalry of the Teutonic and Slavic forces in the Balkans had reached such a degree of intensity as practically to exclude the possibility of a compromise or understanding. The occasion for the outbreak of the great war grew out of the bitter hatred so often engendered by conflicts in which the passionate desire for the assertion of nationality is involved. A hideous outrage, cutting Austria-Hungary to the quick, inspired, very likely, by the spirit of Serbian nationalistic aspirations, occasioned a demand by the Dual Monarchy to be allowed to interfere in the internal affairs of Serbia. But Russia could not afford to be ostentatiously ignored in an intervention by Austria-Hungary in Balkan affairs, or to acquiesce in the reduction of Serbia to a dependent position. This would have destroyed Russian prestige throughout the peninsula, and might have undermined the vital interests of Russia at the Straits. A blow against Serbia had the effect of a blow against Russia.

Europe, as we have seen, was well stocked with masses of explosive or inflammable material, the potential causes of the cataclysm. The antagonism of Slav and Teuton in the Balkans was the agency that applied the torch; international friendships and alliances became the media for the transmission of the destructive forces; and the spirit of militarism, impelled by an inhuman and unnatural conception of honor or professional duty, fanned the flames, until the conflagration became general.

CHRONOLOGICAL TABLE

DATE		PAGE
1552.	Bishopric of Metz annexed by France	16
1581.	Russian occupation of Siberia began	17
1697.	Alsace became French territory	15
1700.	Russia secured the port of Azoff	147
1703.	Peter the Great founded St. Petersburg	18
1711.	Russia forced to give up Azoff	147
1739.	Black Sea closed to Russian Navigation	147
1766.	Lorraine united with France	16
1772.	First partition of Poland	19
1774.	Russia regains freedom of the Black Sea	147
1792.	Russia acquires the site of Odessa	148
	Second partition of Poland	19
1795.	Third partition of Poland	19
1803.	War between Great Britain and France	2
1809.	Russia took Finland from Sweden	18
1815.	Confederation of thirty-eight German states formed	8
	Congress of Vienna	8
	Kingdom of the Netherlands formed	21
	General settlement of European boundaries made	25
1819.	Tariff union established under lead of Prussia	8
1820.	Revolution in Italy	5
1829.	Greece acquires independence	148
1830.	Belgium begins revolt which made her independent of Holland	250
	Virtual annexation of Poland by Russia	20
1831.	Revolution in Italy	5
1832.	Franchise given middle class in England	35
1834.	Zollverein of seventeen German states formed	8
1839.	April 19. Two treaties guaranteeing perpetual neutrality of Belgium signed by five powers in London	251
1840.	First cotton factory built in Russia	56
1848.	Piedmontese army defeated at Custozza	6
	Whirlwind of revolution in Europe	10

DATE		PAGE
1849.	Hungary crushed by the Hapsburgs	22
	Piedmontese army defeated at Novara	6
	Italians defeated at Magenta and Solferino	6
1850.	Triumph of the industrial revolution in Germany	33
1853.	Commodore Perry induces Japan to open ports for commercial relations with rest of world	297
1856.	War in the Crimea between Russia, Great Britain, France, and Piedmont	148
1858.	William I of Prussia takes steps to accomplish German unity	10
1859.	Intervention of Napoleon III in the affairs of Italy	2
	Napoleon III intervenes against Austria on behalf of Italy	6
	January 27. Kaiser William II born	81
1860.	May 11. Garibaldi landed at Trapani	6
	October 1. Neapolitan Bourbons defeated on banks of the Volturno	6
1861.	Austria conciliates Hungary	22
	Bismarck carries out plans for German unity	10
	March 17. Victor Emanuel ascends throne of Kingdom of Italy	6
1863.	Bismarck delivers "Blood and Iron" speech	12
1864.	Regulation of situation of Schleswig and Holstein	12
1866.	War between Prussia and Austria	13
	Battle of Sadowa or Königgrätz	13
	Austrian disaster and exclusion from the German Confederation	23
	Annexation of Venetia to Kingdom of Italy	7
	Austro-Hungarian compromise	23
1867.	Magyar deputies finally consent to attend the Austro-Hungarian parliament	22
	North German Federation formed	13
	Franchise given laborers in England	35
1870.	National school system created in England	35
	Irish land act passed in England	35
	Withdrawal of French garrison from Rome	7
	July 19. Franco-Prussian War began	13
	August 9. North German Federation signs compact with Great Britain confirming the treaty guaranteeing the perpetual neutrality of Belgium	254
	September 4. Second French Republic formed	42
1871.	January 18. William I declared German Emperor at Versailles	14
	May 10. Treaty of peace signed by France and Germany at Frankfort. Alsace and Lorraine ceded to Germany and France agreed to pay indemnity of five billion francs	13

DATE		PAGE
1887.	Ferdinand of Saxe-Coburg elected ruler of the Bulgarians and Roumelians	154
1888.	Germany secures German East Africa	81
	June 15. Kaiser William II ascends the throne	82
	Germans begin building of railways in Asiatic Turkey	161
	German policy substituting German for Polish farmers in Province of Posen adopted	27
1889.	Kaiser William II makes state visit to Constantinople	160
1890.	March 18. Bismarck's resignation as Imperial Chancellor turning point in the history of the modern German Empire	83
	England cedes Heligoland to Germany	100
	Beginning of industrial revolution in Russia	33
	Germany refuses to renew the re-insurance treaty of 1884	32
	Caprivi becomes Chancellor of the German Empire	84
1891.	Building of the Trans-Siberian railway begun	50
1892.	Count Witte becomes Minister of Finance and Commerce in Russia	56
1894.	Hohenlohe becomes Chancellor of the German Empire	84
	Dreyfus case excites France	46
	Japan and China involved in war over Corea	298
1895.	April 17. China makes concessions to Japan in treaty of peace	299
	December 29. Jameson raid occurs in South Africa.	37, 90
1896.	January 3. Kaiser sends congratulatory telegram to Kruger	90
	Massacres of the Armenians occur as a result of Pan-Islamic spirit	160
	Alliance between France and Russia formed	32
1897.	Manufacture of artificial indigo begun in Germany	76
	President Faure visits St. Petersburg and Franco-Russian alliance is announced	85
	September 1. Kaiser's "divine right" speech delivered at Coblentz	85
	November 1. News of murder of two Catholic missionaries in China reaches Berlin	91
	December 15. Kaiser's "mailed fist" speech made at banquet to Prince Henry at Kiel	91
1898.	"Place in the sun" demanded by Count von Bülow	96
	Admiral von Tirpitz appointed chief of Staff of German Navy	97
	Fashoda incident creates tension between France and Great Britain	101, 109
	Roumania, Serbia, and Montenegro recover their independence	148
	Germany demands concession of territory from China in compensation for murder of two missionaries	94

DATE		PAGE
	March 5. Germany secures from China lease of Kiau-Chau which gives her dominant influence in province of Shantung	300
	Kaiser and Kaiserin make state visit to Jerusalem and Kaiser declares the Mohammedans can ever rely on his friendship	161
	Russia secures lease of Kwangto from China and begins fortification of Port Arthur	300
1899.	Kaiser's strong naval policy speech delivered at Hamburg	89
	Germany purchases islands in Pacific from Spain . .	97
	Bill providing for large German navy passed	95
	October. Second Boer war begins	37
	Russia abrogates Finnish constitution	19
1900.	February 8. Bebel, chief of Socialists, predicts that a conflict with England would be greatest calamity that could befall Germany	100
	von Bülow becomes Chancellor of the German Empire .	101
	German laws codified	72
	Boxer uprising occurs in China	300
	Bebel, Socialist leader, opposes in Reichstag proposed navy increase for purpose of war against England .	125
1901.	Turks attempt to secure territory at head of Persian Gulf, but Great Britain interferes	163
1902.	January. Prince of Wales, now George V., visits Berlin on occasion of Kaiser's birthday	105
	January 30. GreatBritain and Japan sign convention of alliance	301
1902.	Polish children punished for refusing to recite the Lord's Prayer in German	27
1903.	King Alexander of Serbia assassinated	205
	Germans secure concession from Turkey for railway to Bagdad	164
	Prussia makes large appropriation for purchasing Polish lands for use of German colonists.	28
1904.	February. War begins between Japan and Russia .	50, 302
	April 8. Anglo-French Convention consummates the *Entente Cordiale*	109
	July 28. Germany takes advantage of Russia's conflict with Japan to force Russia to sign treaty very favorable to German commerce	108
	October. France and Spain sign agreement relative to their interests in Morocco	109
1905.	January 1. General Stoessel surrenders Port Arthur to General Nogi	104, 302
	January. France tries to induce Sultan of Morocco to assent to reforms and the principle of the "Open door"	115

DATE		PAGE
	March 10. Russians defeated by the Japanese at Mukden	115
	March 23. Kaiser's speech at Bremen refers to his dream of world dominion and states "We are the salt of the earth"	112
	The Sultan of Morocco, at suggestion of Germany, calls for a conference of signatories to the Madrid Convention	118
	Great Britain's alleged promise to support France in Morocco claimed by the *Matin*	123
	Distribution of French and British fleets arranged to afford mutual protection	110
	May 27. Russian fleet defeated by Japanese in the Straits of Tsushima	114, 302
	August 12. New treaty of alliance between Japan and Great Britain signed	302
	September 5. Treaty of peace between Japan and Russia signed at Portsmouth, U. S. A.	50, 302
	October 2. Building of the first *Dreadnought* begun	124
	Russia restores to Finland her liberties and privileges	19
	October 30. Tsar of Russia promises people right to participate in legislation	51
	Kaiser visits King Alfonso and goes to Tangier	113, 115
	Delcassé, French Foreign Minister, forced through German intrigue to resign and the Kaiser rewards von Bülow	118
1906.	January 16. Conference to consider Moroccan affairs convenes at Algeciras	120
	Belgian and British military experts study technical problems of British aid to Belgian army in event of German invasion	262
	Universal suffrage of both sexes granted the Finns by Russia	19
	First Duma in Russia convened	51
	British government aids Irish to purchase land	28
	Self government granted the Transvaal	38
	November 10. First *Dreadnought* launched	124
1907.	Universal suffrage granted in Austria	24
	Self government granted Orange River colony	38
	Mulay Hafid rebels against Abdul Aziz in Morocco	133
	Great Britain and Russia settle their differences in Asia by agreement	129
	New treaty between Japan and Russia brings complete harmony of views in respect to China and the Far East	303
1908.	Mulay Hafid recognized by the Powers as Sultan of Morocco	134

DATE		PAGE
	German Bill providing for compulsory expropriation of Polish lands becomes law	28
	Ferdinand of Bulgaria assumes the title of Tsar	171
	Austria-Hungary annexes Bosnia and Herzegovina, and Serbia, under pressure of the Powers, assents	171
	Kaiser in speech at Döberitz claims a desire of other nations to surround and provoke Germany	278
	October. Kaiser in *Daily Telegraph* interview attempts to allay the suspicions of the British	131
	Triple Entente formed by Great Britain, France and Russia, becomes a factor in European diplomacy	130
	Edward VII and Nicholas II have important meeting at Reval	170
1909.	February 9. Germany recognizes by agreement the special interest of France in Morocco	134
	Compulsory potash syndicate or "trust" formed in Germany	72
	July. von Bethmann-Hollweg became Chancellor of the German Empire	84
	Revolution in Turkey forces Abdul-Hamid to resign, and his brother, Mohammed V., becomes Sultan	174
	Congo Free State becomes annexed to Belgium	255
1910.	South African Union formed	38
	Pan German League and Navy League formed in Germany to encourage acquisition of greater colonial empire	125
1911.	An Imperial Parliament proposed for British Empire	38
	Italy begins the war on Turkey which secures Tripoli for her	175
	April. France finds it necessary to send troops to Fez to maintain order there	134
	May 31. Local self government granted Alsace and Lorraine by the German government	14
	July 2. Germany sends gunboat *Panther* to Agadir in Morocco	135
	Germany on verge of war with Great Britain and France	136
	Belgian Ambassador at Berlin informs his government that Belgium's neutrality may be violated	263
	July 21. Lloyd George in speech at London shows that Great Britain is prepared to stand by France	136
	September 18. Stolypin, Russian prime minister assassinated	279
	November 4. Germany and France settle their differences about Morocco by mutual concessions.	136
	November 9. The German Crown Prince in Reichstag applauds bitter speech against England	137
1912.	Russians and Finns given equal rights in Finland	281

DATE		PAGE
	Great Britain and France agree to act together to prevent aggression and to preserve peace	270
	July 30. Yoshihito becomes Emperor of Japan . . .	303
	War in the Balkans brought about by Turkish atrocities.	179
1913.	Germany provides for great increase in her army . .	192
	Coup d'état and revolution in Turkey by Young Turks .	308
	War in the Balkans renewed after peace conference at London fails	185
	August 10. Treaty of peace ending the Balkan war signed at Bucharest	187
	Constitutional government restored to Croatia . . .	200
1914.	Germany completes her remarkable system of strategic railways on Belgian frontier	276
	April. Arms and ammunition made in Germany landed in Ireland for use in revolt	283
	May 25. Home Rule bill finally passed by House of Commons	42
	June 22. Six days before the assassination the Serbian minister at Vienna warns Austrian Government that the Crown Prince's visit to Bosnia would be perilous on account of probable existence of plot . . .	205
	June 28. Crown Prince Franz Ferdinand of Austria assassinated at Sarajevo, capital of Bosnia . . .	202
	July 1. Kaiser formally opens the enlarged Kiel canal .	277
	July 23. Austria-Hungary accuses Serbia of complicity in the murder of the Archduke and demands satisfaction	203
	July 23. Austro-Hungarian government sends ultimatum to Serbian government and demands answer before 6 o'clock of the evening of July 25th . . .	209
	July 24. Austria-Hungary informs the powers that ultimatum has been sent to Serbia, thus leaving only one day for any effort to have matter adjusted . .	213
	July 25. Italy notifies Austria-Hungary that she would regard any occupation of Serbian territory, however temporary, as a violation of the terms of the Triple Alliance	330
	July 26. Home Rule troubles in Ireland culminate in attacks on soldiers, who fire and kill four and wound sixty	284
	July 26. The Kaiser returns to Germany from Norwegian cruise	233
	July 26. Austria-Hungary mobilizes	238
	July 26. British government proposes conference to discuss Serbian affair and France and Italy at once accept	231
	July 27. Russia approves plan of proposed conference .	232

DATE	PAGE
July 27. Germany declines to participate in proposed conference	233
July 28. Austria-Hungary declares war on Serbia	143, 238
July 28. Austrian Ambassador at Berlin informs Sir Edward Goschen "That a general war was most unlikely as Russia neither wanted war nor was in a position to make war"	286
July 29. The Tsar telegraphs the Kaiser that public indignation would force him to protect Serbia against Austrian aggression and asks Kaiser to restrain his ally	237
July 30. The Kaiser telegraphs to the Tsar remonstrating against alleged mobilization of Russian army	240
July 31. Sazonoff, Russian minister of foreign affairs, agrees to maintain waiting attitude if Austria would check advance of her troops on Serbian territory. On Austria's refusal Russian government orders general mobilization	241
August 1. Germany announced a state of war with Russia	143, 242
August 3. Germany declares war against France alleging that aggressive acts have been committed by the latter on German territory	243
August 4. Herr von Jagow replies to remonstrance of British ambassador that Germany had to get into France "by the quickest and easiest way"	267
August 4. Germany announces state of hostilities between herself and Belgium	142
August 4. King of the Belgians appeals to British government to safeguard the integrity of Belgium	256
August 4. Great Britain declares war against Germany	142
August 4. Chancellor von Bethmann-Hollweg acknowledges wrong done to Luxemburg and Belgium and promises reparation	267
August 6. Austria-Hungary declares war against Russia	142
August 6. Serbia declares war on Germany	142
August 7. Montenegro declares war against Austria-Hungary	142
August 12. France declares war against Austria-Hungary	142
August 12. Montenegro declares war against Germany	142
August 12. Great Britain declares war against Austria-Hungary	142
August 16. Japan sends ultimatum to Germany in regard to Eastern affairs	305
August 22. The *Entente* powers guarantee independence and integrity of Turkey if she remains neutral	311
August 23. Japan declares war against Germany	142

INDEX

ADDENDA

Since the publication of the second edition of this volume, my attention has been directed to the views of Rear-Admiral F. E. Chadwick of the United States Navy on the problem of war and peace. These views with some minor variations have appeared successively, in an article in *The American Journal of International Law* for July, 1912, in a pamphlet entitled *The Anglo-German Tension,* in January, 1913, and in the *North American Review* for June, 1915. These articles reveal so keen a grasp of one of the most elementary factors behind the international crisis, the method suggested for the elimination of conflicts is so radical, and the treatment of the theme involves so many important features of the material as treated in this volume, that I desire to present a brief analysis of Admiral Chadwick's principal argument.

The menacing aspect of affairs throughout the world, "the outlook over an angry sea under a gathering tempest," impressed Admiral Chadwick with the futility of most peace programs which deal solely with symptoms or effects and disregard the causes. "True hygiene," he says, "is prevention not cure." He proposes to investigate the basic facts for the lamentable situation which prevails throughout the greater part of the world, and he observes:

"It is almost axiomatic that by far the greatest cause of war is the struggle for special commercial advantage. It is for this that Japan is occupying Manchuria and excluding the rest of us; that France took Algeria, developed a colonial empire a third larger than the United States, and finally swallowed Morocco, despite the convention of Algeciras, and in so doing developed by-products which made the present world-war a certainty. But France is far from being alone. All have struggled for these special spheres of influence, so that in these last thirty-odd years the lands of every race not up to our ideals of efficiency have been taken without so much as a 'by your leave' to the occupants thereof or to the rest of the nations."

Admiral Chadwick selects the extension of French pretensions and influence in Morocco for more detailed examination, probably because it is typical of the general movement and because it involved the greatest peril to the peace of Europe. The Moroccan Question brought Great Britain and Germany to the very brink of hostilities in 1911. The root of such a venomous factor in international relations goes back to a secret agreement between France and Spain for the eventual partitioning of Morocco with Great Britain's compliance. Admiral Chadwick declares that the bargain between these powers for the disposal of Morocco "was the Pandora's box which contained all the woes of which the present situation in Western Europe is the result." Germany refused to permit the accomplishment of this covert design. She could not afford to be officiously disregarded. Because Germany's appearance in the society of the nations was so late she insisted all the more emphatically upon international recognition. Her energetic protest resulted in the Algeciras Conference which established the twofold principle of the integrity of Morocco and of the equal opportunity of all nations in the exploitation of the country. But France violated the convention by invading Morocco and occupying Fez.

Admiral Chadwick believes that the alleged necessity for this intervention was unfounded. Whether the situation really demanded the armed interference of France is a subject of controversy that need not detain us. The consequences of French intervention are manifest. The Agadir incident was Germany's declaration that she considered the Moroccan Question reopened. France was constrained to acquiesce once more in the principle of the "open door" and grant territorial compensation in the Congo region in exchange for the virtual recognition of a French protectorate in Morocco. But the mutual suspicion of Great Britain and Germany had been alarmingly accentuated. "Italy driven by the appropriation of Egypt and of Algiers, Tunis, and Morocco, seized the only remaining chance in Africa and occupied Tripoli at the cost of an expensive war." The Balkan Wars followed immediately and we all know now what dangerous hidden fires of bitterness and resentment they left behind them.

Admiral Chadwick holds the "vicious" system of spheres of influence responsible for the deplorable state of international jealousy and animosity, because in acquiring them the nations have been actuated by a desire to secure exclusive commercial advantages. He predicted in 1912 that if the principle of spheres of influence should hold, "the one nation which most needs an outlet would be prevented by circumstances which can only be overcome by war." The population of Germany is increasing rapidly. "In fifteen years there will be 80,000,000 in Germany on land which will now grow food for but 65,000,000." "Does the great combination," he asks, "which now forms the Triple Entente propose, in order to support their several spheres of influence, to fence in or reduce Germany to nullity? If so, and if as a result of a gigantic war Germany should succumb, Russia would be the great beneficiary. Europe would be dominated by the Slav."

I have myself endeavored to show that the chief aims of German foreign policy were to preserve the "open door," wherever it still existed, and to secure a proportionate share in any new apportionments, or fresh appropriations of colonial territory, and that there is nothing fundamentally unreasonable in such an attitude. While the methods employed by Germany in supporting her foreign policy were not always tactful, her position in the Moroccan controversy both in 1906 and 1911 was logically defensible. Germany, moreover, as Admiral Chadwick very discerningly observes, in insisting on the "open door" in Morocco was serving British interests as well as her own.

Of course the Moroccan Question is not an immediate cause of the present war. It had been settled, presumably, nearly three years before the present whirlwind broke, on the basis of the "open door" and compensation in the Congo.

It can scarcely be denied that Germany exaggerated the significance of opportunities and influence in the distant, undeveloped, semi-barbarous parts of the earth. German sea power was developed ostensibly for the vindication of Germany's claim to a share in the trade and control of these remote lands. But it was an absurdity to regard this ambition as a suitable corner-stone of foreign policy. The profits of all the trade that Germany might hope to acquire in such colonial areas would never pay for the German navy. The most lucrative trade enjoyed by any power with a dependency, a trade which is traditionally regarded as a unique source of riches, is that of the British Islands with British India. But the value of German exports to the British Islands in 1913 was about $400,000,000, while the value of the exports from the British Islands to British

India did not exceed $350,000,000. Carried away, nevertheless, by the commonly exaggerated notions about the value of commercial penetration and influence in distant parts, Germany worked with obdurate energy in building up her formidable fleet to the dread and annoyance of her best customer. A commercial house which pursued similar methods would be bankrupt within a year. But the leaders of public opinion in Germany reject with scorn the analogy between states and commercial enterprises.

These remarks are intended as an indication of the element of fatuity in German naval policy, not as a criticism of Admiral Chadwick's argument. For the opinion certainly prevailed in the influential classes that the utmost exertion ought to be made to secure for Germany her fair share in the opportunities afforded by the outlying parts of the world, and the existence of this opinion is an essential fact among the general causes of international discord.

Admiral Chadwick's observations lead him to the conclusions, (1) that war in a general sense is always the result of unequal opportunity in trade, and (2) that therefore the removal of every artificial restriction upon trade will abolish the causes of war, at least between civilized nations.

He advocates as a remedy for the extirpation of the causes of strife the internationalization for trade purposes of all the regions brought under the control of the different Great Powers since 1880, making the opportunities for trade and exploitation in these regions absolutely equal for all nations, and thus establishing the principle of the "open door" in an area twice as large as the United States. "Segregation in exploitation means wars," he says, "the 'open door' means peace, and if you want peace you must first make war on this great evil of special spheres of interest, just as America is now engaged in war against special interior trusts." The adoption of such a proposal would seem to be no more than an act of elementary justice and compliance with reason. But Admiral Chadwick would not stop here in the effort to realize permanent peace. He would abolish the custom-house absolutely as a commerce preventer. "When men can trade without let or hindrance everywhere," he declares, "just as Maine can trade with California, we shall have taken away all cause for war among civilized nations."

I need hardly state that I agree entirely with the general import of Admiral Chadwick's articles and applaud his proposals. But I believe that the first conclusion as formulated above is too inclusive. Unequal opportunity in trade is unfortunately not the only cause of strife between civilized nations. The Great War has revealed the existence of active sources of conflict which cannot be grouped within the class of commercial causes. Economic factors are undoubtedly the most important, as they are the most prominent, underlying causes. But other motives exist, and some of them involve a more insidious danger, because they are less perceived, at the present time. They are more or less intimately associated with the conception of nationality.

The assertion of nationality and the aspiration for national unity are a vital force in European politics. Irredentism is prolific of international antagonism. It is not primarily by economic considerations that the Italians are instinctively impelled to covet the possession of Trent and Trieste, and the Serbians to aspire to the incorporation with their own kingdom of the Austrian Slavs. It was the collision of nationalities in Macedonia which was the principal cause of the second Balkan War. In all these conflicts, and in many others, the principal animating force has been the spirit of nationality, although commercial considerations may have been a contributory motive.

An abnormal accentuation of the national spirit in the most recent period, spreading from one country to another like an epidemic, rose to a degree of extravagance which was not generally appreciated before the outbreak of the Great War. The term Nationalism is a convenient designation for this phenomenon. Nationalism in this sense is an irrational exaggeration of the personality and pretensions of the state. The Nationalistic creed regards the state as something sacred, self-sufficing, final, existing for its own sake, requiring neither apology nor justification. It regards the service of the state as an act of self-sacrificing devotion to an authority which is rightfully intolerant of all restriction, a power which the individual reason is not qualified to question. Nationalism converts patriotism into a blind, unintelligent fanaticism. Its emphasis on the duty of the citizen to the state is a passionate reaction against the Radical and Socialistic emphasis on the duty of the state to the individual.

In the eyes of Nationalism the ideal aim of the state is power, and therefore the state denies its very nature when it neglects its army. War is a natural function of the state. No self-respecting state will submit to international law or any superior authority against its own opinion and ideals. In the eyes of Nationalism the banishment of war from history would be the banishment of all progress.

The mention of national prestige occurs many times as an uncompromisable factor in the diplomatic correspondence of the days of negotiation before the outbreak of the war. The delicate sensibility of the state had to be jealously guarded against the profanation of even a suggestion of the violation of its prestige or military honor.

Commercial emulation and the competition for markets, by aggravating international differences, have doubtless contributed to the development of Nationalism. The lofty sentiments of disinterested patriotism and honor which Nationalism ostentatiously parades cannot conceal its partial descent from the sordid environment of the market-place. But after ample discount has been made for the element due to commercialism, there is still a force peculiar to itself in Nationalism; while the impulses which animate the Principle of Nationalities are undeniably one of the most important potential causes of the Great War.

In commending his proposal for the general introduction of the "open door" in the spheres of influence, Admiral Chadwick declares: "This arrangement would put an end to rivalry for possession. For why trouble to administer a region in which there is complete freedom of trade?" But as though replying unwittingly to precisely this argument, Professor Hans Delbrück, after admitting that the economic advantages of colonies can be obtained in foreign colonies, if the principle of the "open door" is maintained, and that colonial exploitation of this sort would be free from many national burdens, writes very emphatically that the national, not the economic, consideration should dominate in colonial policy. Trade and commerce, he claims, are only the means for the extension and consolidation of German nationality.

The proposal for the internationalization of dependencies for commercial purposes would not satisfy the aspirations represented by Professor Delbrück. The Germans already enjoy the advantages of such a system in all the British protectorates and spheres of influence. But the unincumbered enjoyment of all the advantages of trade and exploitation in the dependencies of other powers would not be enough. Germany should aim to secure the possession of compact blocks of territory suitable for the expansion of German nationality. Perhaps the unbiased friend of human progress will approve the declaration of

General von Bernhardi that it is a duty to enlarge Germany's colonial empire in the interest of the world's civilization.

There is not merely a divergence of opinion between such views as are represented by Admiral Chadwick and Professor Delbrück; there is an elementary difference in conceptions of political philosophy. The former represents the political doctrine of the Manchester School and Cobden, of whom he says:

"Cobden saw with clear vision when he foretold that freedom of trade would finally be the great factor which would make for peace."

Delbrück stands for principles regarding the nature and function of the state which are diametrically opposed to those of Cobdenism; and Delbrück might reasonably contend that in the colonial aspects of his political philosophy his view is idealistic, while that of Admiral Chadwick is materialistic or utilitarian.

It is a fundamental article of political faith for Professor Delbrück and others who hold the same doctrines that the preservation, strengthening, and perpetuation of nationality, apart from any consideration of practical expediency or material advantage, should be the supreme consideration of the state. Nationality must be cherished with steadfast devotion as a sacred duty to civilization.

It follows that, since there are still aims and ideals of a non-commercial character capable of producing conflicts, purposes which nations are disposed to support by force, the class of causes of warfare arising from commercial disagreement, however important it may be, is not inclusive of all causes of hostility.

Feeling that the obvious disparity in colonial possessions inevitably engenders a sense of unfairness, and bearing in mind the attitude of influential circles in Germany as expressed by Professor Delbrück, I suggested in the conclusions of this volume, page 349, that some method might be devised for mitigating the asperity of the situation, just as legislative provisions have been created in the United Kingdom tending to abate the harshness of distinctions in private fortune. Perhaps the internationalization of commercial advantages suggested by Admiral Chadwick is precisely the suitable instrument for this purpose. It would act automatically. The imposition of the burden of administration without the enjoyment of any exclusive privileges would correspond to the taxation of the unearned increment. It would diminish the appetite for possession, and many protectorates and spheres of influence, like private estates abandoned for the tax titles, might soon become available for the benefit or satisfaction of the powers whose resentment at the evident disparity of the present dispensation is an ever active source of restlessness.

Made in United States
North Haven, CT
04 November 2026

21766805R00300